CAMBRIDGE LIBRARY COLLECTION

Books of enduring scholarly value

Botany and Horticulture

Until the nineteenth century, the investigation of natural phenomena, plants and animals was considered either the preserve of elite scholars or a pastime for the leisured upper classes. As increasing academic rigour and systematisation was brought to the study of 'natural history', its subdisciplines were adopted into university curricula, and learned societies (such as the Royal Horticultural Society, founded in 1804) were established to support research in these areas. A related development was strong enthusiasm for exotic garden plants, which resulted in plant collecting expeditions to every corner of the globe, sometimes with tragic consequences. This series includes accounts of some of those expeditions, detailed reference works on the flora of different regions, and practical advice for amateur and professional gardeners.

The Green-House Companion

The Scottish landscape gardener and prolific horticultural writer John Claudius Loudon (1783–1843) published this manual on the practice of greenhouse or conservatory gardening in 1824. In his preface he remarks on the rise of greenhouses, which have 'now become an appendage to every villa, and to many town residences'. The work is directed at professional gardeners (and their employers) who may not have the requisite knowledge and skills to make the best use of this exciting new space. The first part of the work describes the practicalities of siting and constructing the greenhouse, how it should be heated, with what plants it should be stocked and how these should be cultivated. The second part consists of a catalogue of 'all the green-house and frame plants hitherto in cultivation'. Still valuable, this substantial guide offers modern readers a record of the plants that were then available to the enthusiast.

Cambridge University Press has long been a pioneer in the reissuing of out-of-print titles from its own backlist, producing digital reprints of books that are still sought after by scholars and students but could not be reprinted economically using traditional technology. The Cambridge Library Collection extends this activity to a wider range of books which are still of importance to researchers and professionals, either for the source material they contain, or as landmarks in the history of their academic discipline.

Drawing from the world-renowned collections in the Cambridge University Library and other partner libraries, and guided by the advice of experts in each subject area, Cambridge University Press is using state-of-the-art scanning machines in its own Printing House to capture the content of each book selected for inclusion. The files are processed to give a consistently clear, crisp image, and the books finished to the high quality standard for which the Press is recognised around the world. The latest print-on-demand technology ensures that the books will remain available indefinitely, and that orders for single or multiple copies can quickly be supplied.

The Cambridge Library Collection brings back to life books of enduring scholarly value (including out-of-copyright works originally issued by other publishers) across a wide range of disciplines in the humanities and social sciences and in science and technology.

The Green-House Companion

Comprising a General Course of Green-House
and Conservatory Practice Throughout the Year

JOHN CLAUDIUS LOUDON

CAMBRIDGE
UNIVERSITY PRESS

CAMBRIDGE
UNIVERSITY PRESS

University Printing House, Cambridge, CB2 8BS, United Kingdom

Cambridge University Press is part of the University of Cambridge.
It furthers the University's mission by disseminating knowledge in the pursuit of
education, learning and research at the highest international levels of excellence.

www.cambridge.org
Information on this title: www.cambridge.org/9781108074636

© in this compilation Cambridge University Press 2014

This edition first published 1824
This digitally printed version 2014

ISBN 978-1-108-07463-6 Paperback

CAMELLIA JAPONICA.

Var: 1. *Waratah*.
    2 *Lady Hume's Blush*

Half the natural size

Printed by C.Hullmandel

London. Pub.d May 1. 1824 by Harding & Co.

THE

GREEN-HOUSE COMPANION;

COMPRISING

A GENERAL COURSE OF

GREEN-HOUSE AND CONSERVATORY PRACTICE

THROUGHOUT THE YEAR;

A NATURAL ARRANGEMENT OF ALL THE GREEN-
HOUSE PLANTS IN CULTIVATION;

WITH

A DESCRIPTIVE CATALOGUE OF THE MOST DESIRABLE TO FORM A
COLLECTION, THEIR PROPER SOILS, MODES OF PROPAGATION,
MANAGEMENT, AND REFERENCES TO BOTANICAL WORKS
IN WHICH THEY ARE FIGURED.

ALSO,

THE PROPER TREATMENT OF FLOWERS IN ROOMS,

AND BULBS IN WATER GLASSES.

LONDON:

PRINTED FOR HARDING, TRIPHOOK, AND LEPARD,
FINSBURY SQUARE;
AND JOHN HARDING, ST. JAMES'S STREET.

1824.

PRINTED BY RICHARD TAYLOR, SHOE-LANE.

PREFACE.

A GREEN-HOUSE, which fifty years ago was a luxury not often to be met with, is now become an appendage to every villa, and to many town residences;—not indeed one of the first necessity, but one which is felt to be appropriate and highly desirable, and which mankind recognise as a mark of elegant and refined enjoyment.

The taste for these exotic gardens, indeed, has increased much more rapidly than the skill requisite to manage them to the best advantage,—for the progress of imitation is more rapid than that of knowledge; and hence it is much more common to see a green-house, than to see one filled with a proper selection of plants in high health and beauty.

The management of plants in a green-house requires a higher degree of knowledge, than is called for in the management of the open garden; and though this knowledge is fast extending among the rising generation of gardeners, it is not yet in such abundance as to be general.

The object of the GREEN-HOUSE COMPANION is to supply what is wanting in this respect, not only to gardeners, but to their employers. In composing it we have had in view the twofold object of informing such gardeners as may not have attended much to this branch of their profession; and of enlightening their employers so as they may not only be aware when their servant neglects his duty, or performs it properly or improperly; but may, in the case of employing common labourers in their gardens, be able to direct all the operations themselves. The directions we have given, and the reasons for them, are so ample, and so plain, that no lady or gentleman can be at a loss fully to comprehend them, or discern when they are properly acted on.

Though we have chiefly had in view, in preparing these directions, the villa and town greenhouse; yet we have treated the subject of management so amply, and given so complete a catalogue of green-house plants, with all that it can be desirable to know as to their culture and management, that we have no hesitation in presenting our work as a treatise on green-house culture and management as complete as the present state of things admits of. We have composed it in conformity to the best written authorities, and according to what we have seen in our extensive observation and communications with botanical cultivators and nursery-men, in the neighbourhood of London and elsewhere. The work has also had the advantage of revisal by a nursery-man who is extensively engaged in the culture of green-house and hot-house plants at his commercial establishment in the King's Road. This gentleman's chief business is with exotics, and he supplies and manages town green-houses, and executes London contracts for plants in pots, to a considerable extent: he is therefore peculiarly fitted for the task he so obligingly undertook at the request of the publisher. To mention his name he considers would look like a partial advertisement.

There is one advantage which this work possesses over every other of the kind, which, as it may not strike at first sight, we shall here point out. It is the arrangement of all the green-house plants of this country under the natural orders of the system of Jussieu. In our Introduction to that part of the work which contains this arrangement (see PART II. page 1.) the benefits derivable from it are detailed, and we are here desirous of directing the reader's particular attention to the importance of grouping plants on the green-house stage or open garden platform, according to their natural affinities. Let us not be thought dogmatic in advising every master to insist on his gardener's adopting it as far as circumstances will allow. Whoever understands a

little of the natural system will find few difficul-
ties;—but the prejudices of those who do not,
will start thousands of obstacles.—Still let the
master insist on trying the plan; a thousand dif-
ficulties are often as easily overcome as they are
created,—they lie chiefly in the mind,—and those
minds which are prejudiced must be brought over
by a counter-prejudice, if they are very ignorant;
or by reasoning, if they are a little enlightened.
The idea of natural arrangements is too new to be
generally approved of; but it will soon become as
familiar to the British gardener as the Linnæan
system. Sweet's " Catalogue " has marked out the
road, which will be formed and completed by
Dr. Hooker's "System" and Loudon's " Encyclo-
pædia" of Plants. These works will establish and
confirm our humble efforts : in the meantime we
recommend the master to call in his gardener and
point out to him Part I. chap. II. sect. 6, and the
Introduction to Part II.—let him thoroughly pon-
der what is there said, and afterwards they may
come out to the green-house, where they will find
us ready to show them how the thing may actu-
ally be done.

London, June 24, 1824.

CONTENTS.

b

PART II.

ABBREVIATIONS.

A. H. Andrews's Heaths.
A. R. Andrews's Botanist's Repository.
B. C. Loddige's Botanical Cabinet.
B. M. Botanical Magazine.
B. R. Botanical Register.
S. G. Sweet's Geraniaccæ.

THE

GREEN-HOUSE COMPANION.

INTRODUCTION.

THE most refined enjoyments of society have
gradually arisen from desires more simple, and
even from wants. Man is fond of living beings,
and after assembling those plants around him
which he found necessary for food, he would select
such as were agreeable to the eye, or fragrant to the
smell. A flower in the open parterre, though beautiful and gay, has yet something less endearing,
and is less capable of receiving especial regard,
than a plant in a pot, which thus acquires a sort
of locomotion; and becomes, as it were, thoroughly
domesticated. After choice things were planted
in pots, things rare would be planted in them;
and from things rare to things rare, foreign and

B

tender, the transition would be natural and easy. Tender rare plants in pots would be taken into the house for shelter, and set near the window for light, and hence the origin of the Green-house.

In what age of the world, and in what country a green-house first appeared, it is impossible to determine; it is sufficient to have shown that a taste for this appendage to a dwelling is natural to man; to experience that it adds to his enjoyments; and to feel that it bestows a certain claim to distinction on its possessor. A green-house is in a peculiar degree the care of the female part of a family, and forms an interesting scene of care and recreation to a mother and her daughters, at a season of the year when there is but little inducement to walk in the kitchen-garden, and nothing to do in the parterre or the shrubbery. The progress of vegetation, interesting in all scenes, and at all seasons, is more especially so in a green-house during winter. There the objects are of limited number, brought near the eye by their position, and rendered striking by their contrast with the cold, naked, and dreary scenes which are shut out:—then it is that the genial climate, the life and growth, the deep tone of verdure, and the prevailing stillness of repose within, cause this winter garden to be felt as a luxuriant consecration to man.

3

But all green-houses do not yield the enjoyments which a green-house is calculated to produce; because all are not well contrived, or judiciously managed for that purpose. Some do not know what a green-house will, and what it will not afford; and others expect all its peculiar enjoyments without their accompanying cares. Some erect a green-house of such a form and position, that the plants within can never prosper; others in such a situation, relatively to the house, that if they prosper they can never be enjoyed; and not a few think they have done every thing when they have completed the construction, stocked it with plants, and committed it to the future care of a house servant or mere out-door labourer.

But a green-house, if it is to be worth any thing, must not be trifled with in this manner. It is entirely a work of art: the plants inclosed are in the most artificial situation in which they can be placed, and require constant and unremitting attention to counteract the tendency of that artificial state to destroy them. It is a common notion that a plant in a pot is in a safer state than a plant in the open ground; but this is a most erroneous notion, and directly the reverse of the fact. Placing plants in pots is often more convenient for the cultivator; but it always checks and counteracts the natural habits of the plant: it checks the extension of the roots, and, by consequence, of the shoots ; and it subjects these roots to be alternately deluged by water, and dried up for want of it,—and all this under the best management. Under a careless gardener, if the pots are not properly drained, and this drainage kept in repair, the soil will be kept soaked in water till the roots are rotted ; or by neglecting to shift the plants to pots of a larger size at proper seasons, the roots will get matted so as to derive no benefit from the soil, preclude the water from entering, and thus first stint and then kill the plant. But plants in a green-house are not only in an artificial and injurious state as to the soil, but also as to their climate, and especially as to heat, light, and air. These requisites to vegetation require also to be particularly attended to, so as partly to imitate nature, and partly to effect particular purposes of art on natural principles. Finally, it must be obvious, that where there is so much art, there must be a greater tendency to disease and accident than in ordinary nature ; and consequently, that no small degree of vigilance is required in this respect.

Let none however feel alarmed at these obstacles, or imagine that by a moderate degree of care and attention all these evils may not be avoided, and the enjoyments of the green-house fully obtained. It is the object of this little work to enter into the details of the subject in such a way, as may enable any lady or gentleman, with the assistance of a footman or common labourer, to manage their own green-house as completely as if they employed a regularly bred and skilful gardener.

We shall first offer some suggestions as to the form and construction of green-houses; next, treat of their general management; and lastly, enter into the details of culture applicable to the plants to be grown in them.

CHAPTER I.

Sect. I. *Situation.*

ACCORDING to our ideas of the enjoyments of the green-house, it is essential that it be situated close to the house ; not merely near, but immediately adjoining ; and attached to it either by being placed against it, forming a part of the edifice ; or by means of a corridore, viranda, or some other description of covered passage. The most desirable situation is unquestionably that in which the green-house (fig. 1. Gr.) shall communicate with, and form

Fig. 1.

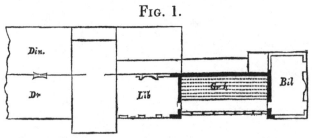

as it were an additional apartment to the library, or breakfast-parlour. If it communicates by spacious glass doors, and the parlour is judiciously furnished with mirrors, and bulbous flowers in water-glasses, the effect will be greatly heightened, and growth, verdure, gay colours, and fragrance, blended with books, sofas, and all the accompaniments of social and polished life.

The next best situation is where the green-house communicates with the drawing-room, to which, by evening illumination, it may lend the same charms as to the morning room.

The next best is where the green-house communicates with the porch, entrance-hall, saloon, or billiard-room (fig. 2. Gr.) ; and in this case, when the entrance to a house is from the north, a noble conservatory or green-house may be projected from the south front, and seen and enjoyed from the windows or glass-doors of the rooms on each side the entrance-hall.

FIG. 2.

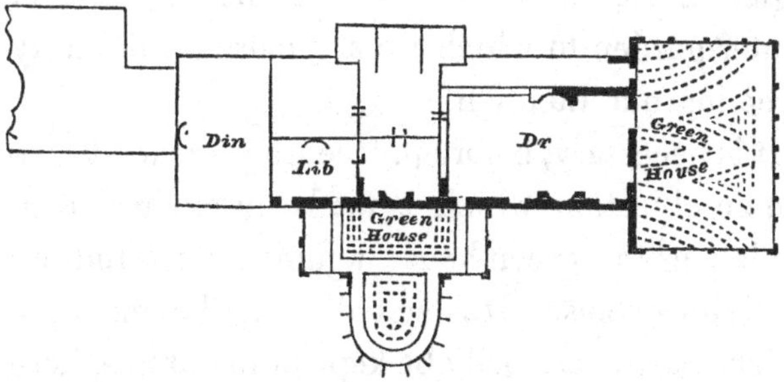

Sometimes the green-house is placed against a wall, or some of the appendage buildings of the house, and is communicated with by a glazed passage from the library or drawing-room; or by a passage inclosed with lattice-work on the south side, a wall on the north, and the roof glazed. Various passages of this sort may be contrived to connect the green-house with the living-rooms of a house ; but when practicable, it is always preferable to have this structure in direct communication with one or more of the principal rooms, so as it may be seen from them as well as walked into. A green-house, however excellent and well managed, if it cannot be seen and entered without going into the open air, can never afford half its appropriate enjoyments during the winter season. Where the green-house is merely a nursery or reserve of plants to supply the flower-stages of the dwelling-rooms, or to grow various plants for the botanist, it signifies little where it is situated in regard to the house ; but such is not the sort of winter garden to which we are more particularly directing our attention.

It is necessary, before proceeding further, to notice an objection which is made by medical men to placing the green-house against the windows of living-rooms. It is stated, that the moisture which must necessarily be kept in the atmosphere of the green-house, in order to ensure the health of the plants, is injurious to the health of animals ;

9

and that this moist atmosphere, every time the windows or glass doors are opened, must necessarily interchange with that of the room. This is unquestionably the case ; and therefore we particularly recommend that no living-room should depend for its ventilation on such of its windows as may communicate with a green-house. This, as may be observed by recurring to the figures (pages 6 and 7), is not the case in the plans which we have given as among the best, and it never need happen in any plan. Where a room is lighted from two sides, one side may look into the green-house without interchange of air, and the other may be devoted to ventilating the room. Where a room is lighted from one side only, and there are two or more windows, one of them may look into the green-house, and the other, or other two, may be used for light and ventilation.

But how are we to get over the difficulty of opening even one of the windows of a room into a green-house, and thereby interchanging more or less the two atmospheres? It must be acknowledged that this difficulty cannot be got completely over, and all the advantages of the communication retained. By double glass doors, and a space between forming a sort of porch, and by having this porch separately ventilated, there would be little danger of interchanging atmospheres : but then the double glasses would greatly impede the view; and having two doors to open instead of one, before entering the green-house, is also a drawback to its enjoyments. The evil, such as it is, must, we fear, be submitted to, especially as it is not a great one; for it is not said that the air of a green-house is unwholesome, but merely that it is more charged with vapour than the common air of a room;—in other words, that a given volume of such air inhaled, will afford less oxygen to the blood than air not so much impregnated with moisture. When it is considered how small a portion of air would be interchanged, and how large a portion of very dry air the small volume of moist air would have to mingle with, the injury to the atmosphere of the room may be considered as very small indeed :—add, that in many cases, and especially where there was a tendency to pulmonary consumption, it would be reckoned favourable rather than otherwise. Wherever a few plants in pots, or bulbs in water-glasses, are kept in a room, the same objection may be raised ; for though the surfaces of the pots and glasses may be completely covered with moss to lessen evaporation, yet that covering is far from being able to prevent it. However, those whose scruples as to health will not allow them to open into their green-houses from a sitting-room, may still look into them from

thence, and, fixing the glazed casements that separate their atmospheres, may enter by some ex-
terior communication to a more intimate enjoyment of the winter garden.

SECT. II. *Position of the Green-house.*

The position of the green-house is the next point
of consideration. The south south-east is unquestionably the best, and corresponds also with
the best position for a breakfasting-room, which
ought to embrace the morning sun, and invite to
go abroad. But this, though preferable both for
the plants, from which it dries up the damps ge-
nerated during the night, and for the illumination
of their foliage and flowers to the spectators in the
living-room, is not the only position. Direct south
is nearly as good ; and east, and from that any
point to south-west, will answer well. A west
aspect has least beauty at that time of the day
when it is most wanted ; it is in the shade all the
morning, and especially in the mornings of winter,
when what is in shade is cold, moist, and unin-
viting. Besides, a green-house with a west or
even south-west aspect, requires much more fuel
to keep it heated, than one in any other aspect.

A green-house is sometimes placed in an angle
or recess of a house, in whose architecture there
are several breaks. This, other circumstances be-

ing favourable, answers very well where the exposition is to the south; but when the aspect is east or west it is worse than any, because during winter the influence of the sun is little felt before eleven, or after two o'clock, and between eleven and two very few of his rays would fall on an eastern or western surface. In short, south, and south south-east, as we have said before, are the only aspects for a green-house to be enjoyed as a winter garden. A green-house, to be supplied with a succession of forced plants from pits and stoves, may be put in any position, and if expense of fuel is no object, may even front the north ; but this is a sort of green-house attended with greater expense than is contemplated by our present views of the subject.

Sect. III. *Form of the Green-house.*

The form of the green-house comes next in order, and depends on various considerations, as to mode of culture, harmony with the architecture of the house, light, heat, air, rain, &c.

There are two modes of cultivating plants in green-houses ; one, and the most usual mode, is, by growing them in pots ; and the other is by growing them in beds or borders of soil, as shrubs or other plants and trees in a shrubbery.   When this last mode is adopted, the magnitude of the structure is required to be larger than for a green-house of pots, and the form must be such as to admit of the roof being removed in the summer season, to give the plants the benefit of the direct influence of the sun and the weather. It is of importance to be impressed with this fact, that unless the roof of a conservatory can be, and is removed every summer, the plants within will soon become naked and unsightly below; nor will any mode of pruning or cutting down prevent this result of the want of air, wind, and the direct influence of the sun. As straight lines, whether in metal or timber, are easier put together, or separated, than crooked or curved lines of any description, it is evident that a parallelogram, or some other right-lined figure, must be the best for the conservatory.

A green-house appended to a dwelling-house should undoubtedly harmonize with that whole of which it is a part. But much less will effect this harmony than what is generally imagined. Few old villas, on a small or moderate scale, display much of design in their masonry or brick-work. With all such, any plain form of glazed structure (fig. 3.) will accord.

Fig. 3.

In more modern houses, where columns, pedi-
ments, cornices, and other architectural finishings,
present themselves, something of the same kind
should enter into the architecture of the green-
house (fig. 4.); but always in a subordinate degree,
and never so as to interfere with the admission of
sufficient light for the health of the plants, or
the power of adequate heating and ventilation.

Fig. 4.

Where a house is characterized by some particular style of architecture, it is easy to impress
that style on the green-house. The form of the
heads of the doors and windows, peculiar to the
different orders of Gothic architecture, can readily

be imitated in the front sashes and doors of a green-house (fig. 5.); and in the case of Grecian architecture, the mouldings of any of the orders are readily applied to the styles, rails and bars, and to the standards and other posts : and even columns may be introduced in very considerable erections.

Fig. 5.

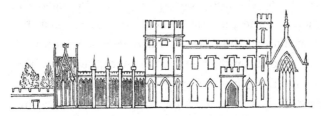

But the grand point which influences the form of a green-house, as it does also its situation and position, is the necessity of admitting the full influence of the sun. For this purpose the roof, on the side next the south at least, and if possible on the east and west sides, must be wholly of glass. It was formerly the custom to form the roofs of green-houses of opaque materials; but this description of architecture, as well as the culture of green-house plants, was then in its infancy. The sickly condition of plants wintered in such of these houses as still exist in this country, affords evidence of their unfitness for the purpose for which they were intended.

As it is found necessary to have the roof and ends of green-houses transparent, so it has been found advantageous to have this transparent tegument of a particular form or inclination of surface relatively to the rays of the sun. It is found that when the sun shines obliquely on any body, whether transparent or opaque, a great many of his rays pass off; and consequently, in the case of opaque bodies, do not communicate their heat to them, and in the case of transparent ones do not penetrate through them. It is therefore desirable to have the glass roof of green-houses as much as possible in a form or slope, which shall form right angles with the sun's rays. But as the sun changes his position in the heavens every day in the year, and every hour in the day, how can this be done? The answer to this (and which was first given by Sir George Mackenzie, a learned horticulturist) is, " Make the surface of your green-house roof parallel to the vaulted surface of the heavens, or to the plane of the sun's orbit :"—in other words, let the roof of your structure be a dome of glass, with only a small part (about one-third of the whole) on the north side opaque. On such a roof the sun's rays will fall in a perpendicular direction in one point, whenever he shines, throughout the whole year. A most elegant house on this principle (fig. 6.)

was erected for the late Lord St. Vincent, by Messrs. W. and D. Bailey, of Holborn, London.

FIG. 6.

But this elegant plan, and all others where the surface is composed of curved lines, is more expensive to execute than where straight lines only are employed; and therefore the question of greatest practical importance is, what form of straightlined surface will answer best as to the admission of the sun's rays? The answer is obvious—a polygon, or many-sided figure, founded on a semidome.

All other circumstances being suitable, the curvilinear form (fig.6.), and polygonal form (fig.7.), are the best for either a green-house or a conservatory; but these forms will not always harmonize with

c

other circumstances, and therefore must frequently give way to such as are more economical or better adapted to particular situations.

FIG. 7.

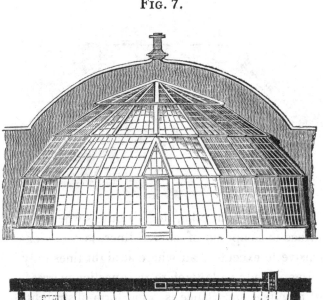

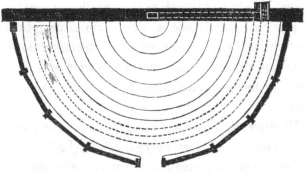

The form of a green-house is, to a certain extent, influenced by the circumstance of its requiring to be heated by art. That form which has the ends and roof opaque, is unquestionably more easily heated than any other; but then there is so great a deficiency of light, as to render such a house of no use in the modern culture of plants. The idea of darkening a house at the ends or from above, for the sake of economising fuel, is therefore quite out of the question; but some varieties of form have a tendency to waste heat, without producing a counterbalancing benefit by light; such as where the upper part of the roof is exposed to the north, and where there are glazed porches of entrance from that quarter. Large windows to the north ought also to be guarded against; and the entrance door of a green-house ought never to be on the north side, unless protected by a close porch or other ante-inclosure.

The form of a green-house must be such as to admit of ventilation, by the entrance and exit of air. In a conservatory we have already stated it as an essential point, that the roof should be removable in summer. It is also a great advantage to that description of winter garden, to have the sashes of the side or sides and ends removable; thereby leaving the plants, in effect, in the open air, and subject to all the weather of the season.

For a winter garden of pots it is not essential that
the roof be removable, because the removal of the
pots to the open air effects the same purpose : but
it is of great consequence that there should be a
power of opening the roof, as well as the south
side and ends, to admit air to enter, circulate, and
escape, every mild day during the winter months.
The common mode of effecting this in right-lined
houses, is by having the roof and front composed
of sashes which slide in grooves, and let down and
open, or draw up and shut, at pleasure ; and the
doors being generally at the ends, the keeping
them open answers for these parts of the structure.
In houses with fixed roofs, and more especially in
curvilinear houses, it is common to have opening
sashes in front, and opaque shutters in the top of
the back wall which open, and by means of these
a current is made to ascend through the house.
In some houses, as in that of Lord St. Vincent already referred to (p. 17), ventilation is effected
solely by shutters in the front wall below the
glass, and others in the back wall close under the
angle of the roof.

It remains only to consider, how far the form
of a green-house is influenced by considerations
relatively to the weather, and especially to rain,
snow, hail, and frost. As to rain, a certain degree
of slope in the roof is necessary to throw it off

when accompanied by wind, and experience has
pointed out 40 and 45 degrees of inclination as
the two extremes. The maximum of that inclination is most favourable also for throwing off hail
when it falls perpendicularly; and preventing
snow from accumulating in large masses, so as to
break the glass by its weight. No slope of roof,
however, will guard against hail accompanied by
wind; all that can be done, where there is much
to be dreaded, is to adopt panes of glass of a small
size and good quality—say not broader than seven
inches, and of Newcastle rather than Greenock
manufacture. To guard against the breakage of
glass by frost, the slope of the roof should not be
less than 45 degrees, and the panes should not
overlap one another more than a quarter of an inch.
By attention to these two points, no water will
ever lodge in the interstices between the panes,
and consequently, there being none to freeze,
there will be no panes broken. It is frequently
alleged that the frost breaks glass by contracting
it, or by contracting the bars in which it is glazed;
but this is an erroneous conclusion, made without
adequate observation. Nothing is more clearly
ascertained, than that the power of frost to break
glass, is nothing without the presence of water in
the interstices: this water, in the process of freez-
ing, expands and separates the two plates of glass

which being unelastic, necessarily break; the fracture generally commencing at the middle of the lower part of the pane where the water, from its gravity, had accumulated in the greatest quantity.

It may be observed, with respect to the angle of 45 degrees recommended as best for the roof in regard to weather, that it is also considered best in all common forms, in respect to the sun's rays. It was that adopted and recommended by Miller, Speechly, Abercrombie, Nicol, and most of our eminent practical gardeners, and, as the Rev. W. Wilkinson has shown, with great reason, as it admits the sun's rays to pass perpendicularly through the roof during the two seasons of the year when they are most wanted; viz. in April for perfecting blossoms, and in autumn for maturing fruits, and ripening young shoots.

SECT. IV.   *Construction of the Green-house.*

Our last subject of consideration in this chapter is the construction of the green-house. Formerly, no one thought of employing any other materials than masonry or brick-work, timber and glass; lately metals, and chiefly iron and copper, have come much into use for every description of plant habitation. For a green-house or conservatory we have no hesitation in giving the preference to metal over wood, as producing a more light, elegant and durable fabric, and admitting of greater latitude of form and dimension, than timber. The metal we prefer is malleable iron, as the cheapest and strongest; and we know of no manufacturers of more taste, science, and experience, than Messrs. Bailey of London, already mentioned. A number of green-houses erected by them in the neighbourhood of London, will amply justify our recommendation.

But though we prefer iron green-houses, we are by no means against the use of wooden ones, which are erected at less expense, and more easily taken to pieces and replaced in the case of a change of plan, or of residence. In the case of large conservatories, indeed, where the roof is moveable, a mixture of timber and iron in the construction (as iron rafters and sash-bars, and timber styles and rails) is preferable to either alone.

SECT. V. *General Remarks as to building or purchasing Green-houses and Conservatories.*

We have now made such remarks on the plans and situations of green-houses, as we think may be perused with advantage by general readers who have it in view to erect or purchase one. We do not think we could go further into details with profit to such as had not already made the subject their study; unless, indeed, we were to write a treatise on hot-house building, which is no more required for our purpose than an elementary treatise on botany. Whoever is desirous of erecting a green-house, will find the cheapest and best method, first, to consult an architect as to the sort of form which will best accord with his mansion; next, a good gardener or nurseryman as to the situation and position; and thirdly, having fixed on these, let him call in a manufacturer of green-houses, and, stating the data given by the architect and gardener, require of him a plan and estimate of the expense. If the party has well considered the subject himself, or has friends in whose opinion he has confidence, then he may dispense with the architect and gardener; but let him not fail to employ a competent tradesman, one accustomed to build green-houses, to regulate their flues, paths, drainage, chimneys, ventilation, &c., and who will be responsible for its answering the purposed ends.

It very frequently happens, in the neighbourhood of large towns, that old or used green-houses are to be purchased, at sales of villas or nursery grounds, at a very low rate. We think it essential to caution the amateur against the purchase of any of these, at least as a green-house to communicate with a house. There is not one green-house in a hundred that will fit two different situations; and if the situation is to be fitted to the green-house, there is little doubt that the result will be what tradesmen call a bungling job. Besides, the green-houses erected by nurserymen are generally of the coarsest materials and workmanship, and quite unfit for being objects of taste or accompaniments of elegance.

Another caution we must give, is against new schemes of heating green-houses and hot-houses by steam, hot air, &c., with which the country seems at present very prolific. Nothing can equal steam for heating an extensive range of hot-houses; but nothing has yet been found which, all circumstances considered, surpasses smoke flues for heating single houses, and especially green-houses, which seldom require the aid of the flues above three months in the year. Beware also of schemes for heating green-houses from parlour or kitchen fires; or any other pretended scientific and economical modes, which are always attended with great expense and risk, which, not being generally understood, are liable to go out of repair; and, if not approved of by the gardener, form a sort of scape-goat for his errors or neglect.

CHAPTER II.

OF STOCKING THE GREEN-HOUSE WITH PLANTS AND TREES.

HAVING completed the erection of the green-house, the next thing is to furnish it with plants : on the judicious selection of these depends much of the future beauty and effect of the winter garden.

With respect to the kind of plants, as the grand object of a green-house is to produce flowers and verdure during winter and spring, those plants should chiefly be selected which are evergreens, and which come into flower during these seasons. It should further be considered, that where a professional gardener is not kept, only the more hardy and easily cultivated of these sorts should be procured ; as it is always more agreeable to see a thriving deep green assemblage of growth and verdure, than a pale sickly collection of rarities.

SECT. I. *Of Fruit-Trees in a Green-house.*

Previously to selecting the green-house plants, properly so called, it may be proper to inquire how far the green-house may be made to supply a few bunches of grapes, or a dozen or two of peaches, as well as the beauty of winter verdure and flowers. To expect much from any green-house in this way would be absurd; because in proportion as space is occupied and light excluded by the one sort of production, the room for the other must necessarily be diminished : but happily the grape-vine is of such a hardy and accommodating nature, that, by a little contrivance, one or two plants of it may be cultivated in every green-house. The simplest and best mode of introducing vines into a green-house is by planting them outside of the south front, close under the upright glass, and introducing the shoots through an opening in the masonry, or, what is better, in the corner of a movable sash. The vine so introduced may then be trained in a direction from the front to the back of the house, at about a foot distance from the glass; and either on the under edge of the wooden rafter in which the sashes slide, or on a stout wire, or couple of wires, suspended from it, or by any other means, placed parallel to and within 18 inches of the roof. In a green-house where the plants are to be kept in pots, vines may be introduced in this way every four or five feet throughout the length of the front, so that a house 30 feet long would give seven or eight vines; and seven or eight vines so situated and properly managed, will on an average of years produce at least 20 bunches of grapes each, or, in all, 250 or 300 lbs. of grapes.

In a green-house where the plants are to be grown in beds and borders, and the roof of the house removed during summer, it will be of no use planting vines; as the fruit would never ripen, and indeed the shade that would be thrown on the plants below, by vines or other creepers on the rafters, would be so injurious to them as to condemn the practice, even if in other respects advantageous. In a green-house of pots, the plants being set out of doors during summer, the shade of the vines does no injury whatever to the few tender annuals, as balsams, &c., which at that season may be set on the stage as decorations.

A green-house of pots therefore has this advantage over a conservatory, that it will afford a few grapes. By introducing peach trees, figs, and other trees or fruit-bearing shrubs in pots, a further variety of fruit may be obtained ; but, with the exception of peaches and figs, we cannot recommend an attempt to go further, as the space occupied by such articles would be too great ; would interfere with the character and beauty of the house, by presenting a number of naked and common forms in large pots of earth ; and after all, supposing the success as complete as could be expected, the result is only a few common fruits a week or two before they ripen in the open air. There are some, however, who prefer a green-house of fruit trees in pots, to a green-house of verdant varieties barren as to the palate ; but such a taste cannot be considered as elegant or refined.

Peach trees are sometimes planted in green-houses, and trained up the rafters or on a couple of wires in the manner of vines. This may be tried with one or two rafters; but it will be found that the peach tree soon gets too woody and un-wieldy for this mode of training, and also that its blossoms and leaves expand so early as to be in-jurious to the green-house plants by their shade. The least injurious mode, where the plan and circumstances of the house are suitable, is to plant a peach tree at one end, and a fig tree at the other; placing them in the north angle, and training them in part on the end glass, and in part on the back wall where it is not covered by the stage of pots. Or, instead of training them either against the wall or the glass, the trees may be allowed to spread their heads as standards, in which state, being pruned so as not to get crowded, they will be found to bear very well.

The pomegranate and the olive, and also the jujube and carob, might no doubt be fruited in a green-house, as they all produce abundantly in the gardens and fields about Nice and Genoa; but unless where the green-house was very large, or there was a peculiar taste in the owner for growing fruits, it would not be worth while to attempt it. Those who wish to make trial, should adopt large well-drained pots or boxes, or plant in a border against the back wall or ends of the house.

Having limited the fruit trees to be introduced into a green-house to a few vines, a fig, and a peach, we shall give the names of the sorts of those which we consider the most suitable.

Grapes.—Of black sorts the *July* is the earliest; it has a small loose bunch aud round sweetflavoured berry; it bears well, and the leaves are not large.

The *black sweetwater* has a larger berry, closer bunch, and richer flavour; but it is not so hardy or so good a bearer.

The *black cluster*, of which there are several varieties, is a very hardy plant and a great bearer; the berries are small, round, and close in the bunch; but their flavour is strong, and more suitable for making wine than for the dessert.

The *esperione* is an excellent black grape, hardy, a great bearer, round berries, close bunch, and flavour resembling the Hamburgh.

The *black prince* and *black frontiniac* are both good grapes, and should be planted in the warmest part of the green-house.

Of white grapes none are more suitable than the *muscadine*, especially the small early variety ; and next, the common *sweet-water*, and *white frontiniac* ; the latter placed in the warmest part of the house.

In planting these grapes, a white and red sort, or, as the reds or blacks are generally preferred to the whites, two reds and one white, should be planted alternately, in order that there may be a finer effect produced by the fruit when it begins to ripen, and the leaves in autumn when they begin to decay. At that season the leaves of the red sorts take a scarlet, and those of the whites a yellowish hue.

The age of the plants may be two years from the eye or cutting. After planting they should be cut down to the lower edge of the lowest pane of front glass, and the leading shoot trained from that point. This shoot should be shortened every year at the winter pruning, to three or four buds, in order to force it to throw out side-shoots, which must be spurred in, that is, all cut off excepting two or three buds, this being the mode of bearing most suitable for grape-vines so circumstanced.

Particular attention must be paid to the soil in

which the vines are planted, as unless this is of a good quality, and laid perfectly dry below, all other labours with this fruit tree will be in vain. No tree whatever is so much injured by a springy, clayey subsoil as the vine : it is not sufficient, in the case of such soils, to form a drain round the house, or round the border, as that will not prevent the cold moisture from ascending into the superstratum; this superstratum of good soil must be separated from the subsoil by a layer of stones, brickbats, gravel, and lime rubbish, well mixed together and beaten, or rammed into a compact body. On this bottom a compost of loam, dung, and a little sand and lime rubbish should be laid, to the thickness of two feet at least. It is of much less importance what the component parts of this compost are, than that the whole should be laid dry. Four parts of loamy turf from any common or old pasture ; a fifth part of rotten dung, blood, night soil, bones, or any dung no matter how strong or coarse it may be ; and a sixth part of lime rubbish and coarse sand or gravel mixed together, will form an excellent compost, and if there are any bones in it, it will require no renewal or enrichment for many years.

The same compost with an addition of sand and lime rubbish, will answer well for the fig,

olive, pomegranate, and jujube, if a tree of either
of these sorts should be planted in a corner; and
for the peach, the turfy loam, with a small addition of rotten dung and sand, will be sufficient, as
this tree does not thrive in a very rich soil, or in
one where much lime is present. The sorts of PEACHES we would recommend as
standards for a green house, are the *royal george*
and *red magdalen;* and of FIGS the *early blue*, or
blue Ischia and *early white*.

Having introduced such fruit trees as are admissible in the green-house, our next business is
to indicate the plants which shall constitute the
main stock, or the green-house plants properly so
called. With the exception of a few climbers or
twiners, all of these are grown in pots.

SECT. II. *Climbers and Twiners.*

The climbers and twiners introduced into a
green-house should be very few, because, as they
are generally evergreens, they shade the plants
below during the whole winter and spring. Some
introduce them under the roof, in lieu of the vines
which we have recommended, and where there is
a separate house for vines, this is doubtless the
best mode. In a small green-house, however, it
is better to place them in subordinate situations,

as at the ends, or against such props or pillars as may be introduced to support the roof. It sometimes happens also, that they may be grown on iron rods fixed to the stage, or forming arches over the path, or a sort of arbour or porch immediately within the doors of entrance. But whereever they are placed, the great object is to take care that they do as little injury as possible by their shade; and, as it is very difficult to avoid this, the fewer that are planted the better. Many of the sorts may be grown in pots, and trained round a cone of rods or wires fixed to an iron basement or saucer on which the pot stands.

The climbing and twining green-house plants which we would recommend are the following:

Hoya carnosa, B. M. 788, which has red and white flowers in June, and the odour of honey.

Glycine sinensis, B. M. 2083, with fine blueish pea flowers during the same month.

Bignonia grandiflora, B. M. 1398, B. R. 418, with large scarlet trumpet-shaped flowers in July.

Cobœa scandens, B. M. 851, which has purple flowers, and is in bloom all the summer. Great care must be taken to keep it within bounds, otherwise it will overrun every thing in the course of one season. It has been known to grow at the rate of a foot a day, for upwards of two months together.

Convolvulus canariensis, B. M. 1228, is a herbaceous plant with a twining stem, and very showy reddish blue and white flowers which appear in July.

Jasminum grandiflorum, B. R. 91, has a fine sweet scented white flower, and well deserves a place, as it will perfume the house during a whole summer. Some other jasmines which are also climbers may be grown in pots.

Passiflora cærulea racemosa, B. C. 573, and the other hybrids originated from these two species, deserve introduction where there is room. They flower late in the season, and often continue in flower great part of the winter.

Lonicera flava, B. M. 1318, has yellow odoriferous flowers, and merits a place next to the two last genera.

A soil composed of loam and sand with a little peat earth will suit all these plants. It must be well drained by a bed of gravel or rubbish below, as from the water which drops from the plants in pots on the stage, and that used to syringe the house, or to wash the pots, climbers are very apt to be overwatered, and the earth to become sodden; and when this is the case, they cease to thrive.

SECT. III. *Green-house Plants in Pots.*

In recommending a stock of green-house plants

D 2

for pots, we shall endeavour to avoid the common
and very prevalent error of choosing a great number of different species, with more regard to their
variety and value to the botanist, than their
beauty to the general observer or amateur, and
their fragrance and easy culture. We have no
hesitation in stating, that a more beautiful and
fragrant display may be made from a judicious
selection of *heaths, geraniums, myrtles, oranges,
camellias, proteas, salvia, polygala, diosma, gnidia, acacia, melaleuca, nerium, fuchsia,* and perhaps one or two others, than from all the rest
of the green-house plants known to botanists. The
heaths, the *camellias,* and above all the *geraniums,* are an inexhaustible fund of beauty, and
that for every month in the year. The *camellias*
are in perfection at Christmas, and last till the
beginning of February, when they are succeeded
by the *geraniums* and *heaths,* which keep flowering till the succeeding November.

We wish we could impress sufficiently on our
readers, the importance of selecting a few choice
sorts, rather than aiming at a great number of
species, or what gardeners call a greater variety.
The truth is, that within the last fifty years the
accession to our stock of exotics has been so
great, that gardeners are quite bewildered among
them, and the nurserymen at present, in their

recommendation of plants, act as if every pur-
chaser were a botanist. This is the reason why
we see so very few green-houses that present a
gay assemblage of luxuriant verdure and blossoms : on the contrary, they are generally filled
with sickly naked plants in peat soil, with hard
names, which one-half of people of taste and
fashion, and nine-tenths of mankind in general,
care nothing about.

We shall now enumerate such species of the above
and a few other genera as we would reco mend;
and we shall in a future chapter enumerate all the
genera of green-house plants known, and state
something of their characters as to beauty and
cultivation. But we would previously premise
two things : first, that we do not profess to have
included *all* the handsome species of green-house
plants in our list of stock ; and secondly, that we
would recommend to any person wishing to stock
a green-house, not to carry a list to a nurseryman,
but to describe generally the sort of collection
wanted, and leave it to him to supply such species
as will produce it; for some of the species in books
are not known to practical men, rare, dear, have
their names changed, or have been lost; and others
are supplanted by superior species or varieties
more recently introduced or originated. The true
mode, therefore, not to meet with disappointment,

is to employ a respectable nurseryman, who will
not only look to giving present satisfaction, but
to ensuring future favours. Of such nurserymen
there are many round London.

Subsect. 1. CAMELLIAS.

The CAMELLIA is the genus we shall commence
with, as flowering the earliest. The *C. bohea* and
viridis are the plants the leaves of which compose
the black and green teas of China, mixed, as it is
said, with those of *C. sasanqua.* The ornamental
species is the *C. japonica,* an evergreen shrub
which grows to the size of a low tree in China;
it has dark green shining ovate leaves on short
footstalks; and flowers red, white, striped, variegated, semi-double, and double, of various forms,
and without smell. The principal varieties are :

Of *Red,* the single, semi-double, double pale,
dark, large, pæony, pompone, long-leaved,
Greville's, Campbell's, coronet, Middlemist's,
Loddiges', Waratah, and others.

Of *Whites,* the single, semi-double, double,
Willbank's, fimbriated, spotted-leaved, &c.

Of *Red* and *White,* the striped, single, semi-
double, and double; the variegated spotted-leaved,
and some other seedlings not yet named.

Some of these sorts are figured in the Botanical Magazine, in Andrews's Botanist's Repository, in the Botanic Register, in Loddiges' Botanical Cabinet, and references will be found to them in our general catalogue, the last chapter, art. *Camellia*.

New varieties are continually originating, by the nurserymen and other growers, from seeds. A number of hybrids are in an advanced state, but have not yet flowered.

The double red, pale red, pæony, pompone, Greville and Waratah reds; the double whites, and the double-striped, form a very handsome collection.

The soil for the *Camellia* is loam two parts, and sand and leaf-mould or peat-earth one part: they will grow well in loam alone, provided the pots be so well drained as not to retain water. The single red is propagated by cuttings and layers, on which the other species are propagated by grafting or inarching. The *Camellia* is very hardy, and requires little more than 45 or 48 degrees of Fahrenheit when in flower; but when the shoots are growing, which is commonly a month or six weeks after the flowers have faded, they should have more heat by 8 or 10 degrees, and plenty of water both at root and over the top. As soon as they have done growing they should be kept cool to harden the wood, and they may be placed in the open air, in a sheltered and rather shady situation, from the end of May to October.

Subsect. 2. HEATHS.

The *Erica* comes next in order : some of its species are in flower all the winter, a few regularly blossom in March, more in April, and a great variety in every month to November inclusive. We shall give the names of some of the most beautiful for each month.

January to March inclusive, but chiefly in March.

Erica ardens, B. R. 115, flowers red, open, roundish ; plant not exceeding 6 inches high. B. R. 115. B. C. 47.

E. oppositifolia, flower purple, open, ventricose ; height 6 or 7 inches.

E. pallens, pill-shaped, yellow flower; plant nearly a foot high.

E. barbata, with pill-shaped flowers of a white colour, and the plant 10 inches high. There are three varieties, *E. b. major, minor,* and *discolor.* B. C. 124. A. H. vol. 2.

E. vernix, roundish greenish flower, and plant nearly 2 feet high. There is a variety called *E. v. major,* full 2 feet high. B. M. 1139.

April.

Erica mutabilis, purple ventricose flowers; plant
nearly 6 inches high. B. C. 46. A. H. vol. 3.

E. acutangula, white flowers; low plant.

E. finitiana, open purple flowers; plant nearly a
foot high.

E. gracilis, purple pill-shaped flowers; plant 10
inches high.

E. præcox, purple pill-shaped flowers; plant 8
inches high. B. C. 244.

E. baccans; flowers round, purple; plant nearly a
foot and a half high. B. M. 358.

E. aristata ; flowers white, open, ventricose; plant
from 12 to 18 inches high. B.M.1249. B.C.73.
A. H. vol. 3.

E. linnæoides, tube-shaped purple flowers; plant
nearly 2 feet high.

E. sessilifolia, tube-shaped yellow flowers; plant
20 inches high.

E. gelida, ventricose, white, wax-looking flowers;
plant above 18 inches high. B. C. 699. A. H.
vol. 2.

E. costata, tube-shaped red flowers; plant above
2 feet high. A. H. vol. 1.

E. spicata, tube-shaped yellow flowers; plant
nearly 30 inches high. A. H. vol. 1.

E. discolor, tube-shaped white flowers; plant
from 2 feet to 30 inches high. A. H. vol. 1.

May.

Erica fimbriata, pill-shaped red flowers; plant under 6 inches high.

E. squarrosa, ventricose white flowers ; plant not more than 6 inches high.

E. kalmiflora, open red flowers ; plant from 6 to 12 inches high.

E. trivialis, pill-shaped red flowers ; plant 9 inches high.

E. mundula rubra, ventricose red flowers ; plant nearly a foot high. B. C. 668.

E. venusta, and *squarrosa carnea*, elegant tube-shaped red flowers ; plants nearly one foot high.

E. exserta and *mellifera*, pill-shaped purple flowers ; plants nearly one foot high.

E. fragrans, A. H. vol. 2, B. C. 288, *mollis* and *puerilis*, ventricose purple flowers ; plants nearly one foot high.

E. campanulata, A. H. vol. 1, B. C. 184, and *lacticolor*, roundish yellow flowers ; plants from 6 to 12 inches high.

E. arctata, cordata, A. H. vol. 3, *stellata* and *densa*, elegant pill-shaped white flowers ; plants nearly 10 inches high.

E. biflora, B. C. 683, and *ursina*, roundish white flowers ; plants from 6 to 12 inches high.

Erica acuminata, tube-shaped red flowers; the plant from 12 to 18 inches high. B. C. 216.

E. persoluta rubra and *p. conferta,* A. H. vol. 2, pill-shaped red flowers, and handsome plants from 12 to 18 inches high.

E. patens, A. H. vol. 3, *plumosa* and *racemosa,* open purple flowers; plants from 12 to 18 inches high.

E. muscaria, A. H. vol. 1, and *tenuiflora,* A. H. vol. 3, tube-shaped and yellow flowers; plants from 12 to 18 inches high.

E. actea, donnia, persoluta alba and *pinifolia discolor,* A. H. vol. 1, pill-shaped white flowers; plants above 12 inches high.

E. ruffa, versicolor, A. H. vol. 1, B. C. 208, *tubiflora* and *spuria pallida,* tube-shaped red flowers; plants from 20 to 24 inches high.

E. andromedæflora, B. M. 1250, B. C. 521, and *racemiflora,* pill-shaped red flowers; plants nearly 2 feet high.

E. spuria, A. H. vol. 1, tube-shaped purple flowers; plant 20 inches high.

E. ignescens and *hybrida,* A. H. vol. 2, roundish yellow flowers; plants from 18 inches to 2 feet high.

E. Patersoni major, A. H. vol. 1, elegant tube-shaped white flowers; plant 2 feet high, with long trembling foliage.

E. conferta, A. H. vol. 2, and *flexuosa,* A. H. vol. 1,

pill-shaped white flowers; plants above 20 inches high.

Erica enneaphylla, tube-shaped yellow flowers; plant nearly 2½ feet high.

E. viscaria, B. C. 726, roundish red flowers on a plant which grows to the height of 3 feet.

E. concinna, tube-shaped purple flowers; plant from 30 to 36 inches high.

June.

E. longipedunculata, B. C. 103, and *saturejæfolia*, open and ventricose red flowers on plants of the most diminutive size.

E. Banksia purpurea, tube-shaped purple flowers; plant seldom above 6 inches high.

E. droseroides and *elegans*, B. C. 185, B. M. 966, open purple flowers on low plants.

E. Banksia lutea, tube-shaped yellow flowers; plant under 6 inches high.

E. acuta, tube-shaped white flowers; plant seldom more than 6 inches high. A. H. vol. 2.

E. petiolata, A. H. vol. 3, and *primuloides*, B. C. 715, B. M. 1548, roundish white flowers on low plants.

E. bracteata, *pistillaris* and *hyacinthoides*, A. H. vol. 3, roundish ventricose red flowers; plants under a foot high.

Erica blanda, B. C. 13, A. H. vol. 3, and *depressa*, A. H. vol. 2, tube-shaped purple flowers; plants from 6 to 12 inches high.

E. concava, B. C. 134, *congesta*, *Nivenia*, A. H. vol. 2, and *nobilis*, open ventricose purple flowers; plants from 6 inches to one foot in height.

E. halicacaba, A. H. vol. 2, ventricose yellow flowers; plant about a foot high.

E. nivea and *rostella*, roundish white flowers; plants under a foot high.

E. Linnæa superba, elegant tube-shaped red flowers; plant rising 18 inches high.

E. empetroides, lævis rubra, and *margaritacea incarnata*, A. H. vol. 2, roundish red flowers; plants from one foot to one foot and a half high.

E. empetrifolia, B. M. 447, pill-shaped purple flowers; plant rising 18 inches.

E. flammea, A. H. vol. 2, and *erecta*, tube-shaped yellow flowers; plants from 12 to 18 inches high.

E. epistomia, ventricose yellow flowers; plant upwards of a foot high.

E. Linnæa, tube-shaped flowers; plant 18 inches high. B. C. 102. A. H. vol. 2.

E. margaritacea, A. H. vol. 1, pearl-shaped white flowers; plant 18 or 20 inches high.

E. lævis, melanthera, pyroliflora, regerminans and *triflora*; roundish white flowers on plants from 12 to 18 inches high.

Erica splendens, tube-shaped red flowers; plant 18 inches high.

E. reflexa rubra, roundish red flowers; plant nearly 2 feet high.

E. simplicifolia, tube-shaped yellow flowers on a plant from 20 to 24 inches high.

E. absinthoides, aristata serotina, A. H. vol. 3, and *reflexa alba,* roundish white flowers on plants from 18 inches to 2 feet high.

July.

E. paniculata and *lachnœa rubra,* open roundish red flowers; plants under 6 inches high.

E. droseroides major, roundish purple flowers on very low plants.

E. Kennedia and *Parmentiera,* B. C. 197, tube-shaped red flowers; plants from 9 inches to a foot in height.

E. Dicksonia rubra, articulata, comosa conferta et rubra, ventricosa, B. C. 431, B. M. 350, and *hirta,* roundish red flowers on plants from 6 inches to a foot in height.

E. campestris and *protrudens,* purple roundish flowers on plants of a foot high.

E. cistifolia, lachnœa and *rupestris,* roundish white flowers on very low plants.

E. Sparmannia, A. H. vol. 3, tube-shaped yellow flowers; plant a foot high.

Erica lutea, A. H. vol. 1, B. C. 64, and *magnifica*, roundish yellow flowers ; plants from 6 to 12 inches high.

E. Massonia ferruginea and *minor*, B. M. 356, tube-shaped green flowers ; plants nearly a foot high.

E. campanulata, A. H. vol. 1, B. C. 184, *Dicksonia alba, glomerata, Peziza,* B. C. 265, *retorta,* B. M. 362, *brevifolia, rotundifolia, sexfaria,* A. H. vol. 2, and *thymifolia,* A. H. vol. 2, roundish white flowers; plants rising a foot high.

E. inflata, muscosa, pedunculata, ramentacea, A. H. vol. 1, B. C. 446, *Walkeria superba,* A. H. vol. 1, B. C. 256, *canescens,* A. H. vol. 2, *incana, incarnata major, juliana, moschata,* B. C. 614, and *propendens,* B. C. 63, A. H. vol. 2, roundish ventricose red flowers ; plants from 12 to 18 inches high.

E. perspicua, tube-shaped purple flowers on a plant 15 inches high.

E. obliqua, A. H. vol. 1, *pubescens major,* B. C. 167, *Petiveria major et minor,* and *recurvata,* roundish purple flowers on plants from 12 to 18 inches high.

E. exsurgens coccinea and *fulgida,* tube-shaped copper-coloured flowers ; plants 18 inches high.

E. petiveria aurantia, Sebana minor, tetragona and

Thunbergia, B. C. 277, B. M. 1214, roundish orange-coloured flowers; plants from 15 to 19 inches high.

Erica exsurgens, A. H. vol. 1, *decumbens*, *capitata*, A. H. vol. 1, *Humea*, B. C. 389, and *urceolaris*, roundish flowers of a white colour on plants from 12 to 18 inches high.

E. penicillata rubra, *hirta* and *Aitonia*, B. C. 144, B. M. 429, tube-shaped red flowers; plants from 18 to 24 inches high.

E. colata, *pinnifolia coccinea*, *hispida* and *imbecilla*, roundish red flowers; plants from 1 to 2 feet in height.

E. Sebana, A. H. vol. 1, B. C. 23, tube-shaped purple flowers; plant rising 20 inches.

E. decora, A. H. vol. 3, and *glauca*, B. M. 580, roundish purple flowers; plants from 1 to 2 feet in height.

E. mollissima, *Hibbertia* and *Hibbertia minor*, A. H. vol. 3, roundish yellow flowers; plants from 1 to 2 feet high.

E. foliosa, *formosa*, *glabra*, and *Sebana fusca*, *aurantia et lutea*, tube-shaped yellow flowers; plants from 20 inches to 2 feet in height.

E. marifolia, A. H. vol. 1, *Monsonia minor*, *penicillata*, A. H. vol. 2, *pinifolia* and *triumphans*, roundish white flowers; plants from 1 to 2 feet high.

49

Erica melastoma, A. H. vol. 1, tube-shaped variegated flowers; plant rising 2 feet high.
E. conspicua, roundish yellow flowers; plant
nearly 30 inches in height. A. H. vol. 2.
E. rosea, A. H. vol. 2, and *vestita carnea,* tube-
like red flowers on elegant plants rising 2½ feet
high.
E. grandiflora, long tubular yellow flowers; plant
from 30 inches to 3 feet in height. B. M. 189,
B. C. 498.
E. coronata, tube-shaped green flowers; plant upwards of 30 inches in height.
E. vestita alba, tube-shaped white flowers on a
plant 3 feet high. B. C. 243.

August.

E. pumila, roundish red flowers; plant under
6 inches in height.
E. intertexta, ventricose white flowers; plant of
diminutive growth.
E. calycina major, B. C. 594, *obcordata rubra,*
paradisiaca, peltata, flagelliformis, taxifolia,
deflexa and *Noisettia,* roundish or pill-shaped
red flowers on plants from 6 inches to 1 foot in
height.
E. scariosa, roundish purple flowers on a plant a
foot high. B. C. 477.

E

Erica sordida, tube-shaped yellow flowers; plant
10 inches high. A. H. vol. 1.

E. *metulæflora*, ventricose yellow flowers on a
plant from 6 to 12 inches high. B. M. 612.

E. *infundibuliformis*, elegant tube-shaped open
flowers, of a pure white ; plant a foot or more
in height. B. C. 589.

E. *aggregata, densa, Lambertia*, B. C. 3, A. H.
vol. 2, *proboscidea* and *struthiolæflora*, roundish
white flowers on plants from 6 to 12 inches high.

E. *ignescens*, A. H. vol. 2, *Sebana sanguinea*, B. C.
86, and *rigida*, tube-shaped red flowers on
plants from 12 to 18 inches in height.

E. *rubella*, B. C. 658, B. M. 2165, *rubens*, B. C.
557, *Templea, denticulata rubra, muscoides* and
brunioides, roundish red flowers on plants from
12 to 18 inches high.

E. *lucida*, A. H. vol. 2, *quadriflora* and *strigosa*,
roundish purple flowers on plants from 14 to 20
inches in height.

E. *albens*, B. C. 95, B. M. 440, and *denticulata*,
roundish ventricose yellow flowers on plants
from 12 to 18 inches high.

E. *gemmifera*, tube-shaped green flowers; plant
from 12 to 18 inches high. B. C. 457, B. M.
2266.

E. *bandonia*, ventricose green flowers ; plant
nearly 18 inches high.

Erica argentiflora, tube-shaped silvery white flowers; plant from 12 to 15 inches in height.

E. aspera, caffra, B. C. 196, *daphnæflora*, B. C. 543, *fastigiata*, A. H. vol. 1, B. C. 207, *flaccida, pendula, phyllicoides* and *shannonia*, B. C. 168, roundish ventricose white flowers on plants under 18 inches in height.

E. caffra spicata and *tricolor minor*, pill-shaped and longish variegated flowers on plants 18 inches high.

E. carinata, vestita incarnata, verticillata, B. C. 145, A. H. vol. 1, and *mammosa alba*, tube-shaped red flowers on plants generally about 2 feet high.

E. ampullacea, B. C. 508, B. M. 303, *Broadlyana, Salisburia, Jasminiflora* and *pellucida*, B. C. 276, A. H. vol. 3, roundish red flowers on plants from 18 to 24 inches high.

E. mammosa, B. C. 125, A. H. vol. 1, and *mammosa major*, tube-shaped purple flowers on plants from 18 to 24 inches high.

E. pallida and *nitida*, roundish purple flowers on plants 20 inches high.

E. horrida, a tube-shaped dark yellow flower; plant 24 inches high.

E. nudiflora, a roundish yellow flower; plant 20 inches high.

Erica Swainsonia, a tube-shaped green flower ; plant 2 feet high.

E. mirabilis, m. major and *pilosa*, B. C. 606, tube-shaped white flowers on plants from 18 to 24 inches in height.

E. arbutiflora, cupressina and *physodes*, B. C. 223, B. M. 443, roundish ventricose white flowers on plants from 18 to 24 inches in height.

E. tricolor major, variegated roundish flowers ; plants 2 feet high.

E. elongata, tube-shaped red flowers ; plant about 2 feet high. B. C. 738.

E. imperialis, tube-shaped green flowers ; plant nearly 30 inches high.

E. bucciniformis, tube-shaped red flowers ; plant 30 inches high.

E. elata, tube-shaped yellow flowers ; plant 30 inches high, and upwards. A. H. vol. 2.

E. Massonia, greenish white ventricose flowers ; plant upwards of 30 inches high. B. M. 356.

September.

E. calycina minor, B. C. 594, and *tenuifolia*, A. H. vol. 3, pill-shaped and roundish flowers of a deep red on plants of very low growth.

53

Erica Smithia and *Archeria*, A. H. vol. 2, ventricose purple flowers on plants 6 inches high.

E. speciosa, roundish yellow flowers on plants
about 6 inches high. B. C. 575, A. H. vol. 2.

E. Plukenetiana nana, tube-shaped red flowers ;
plant rising 9 inches in height. A. H. vol. 1.

E. declinata, fibula, floribunda, B. C. 176, *cernua*
and *imbricata*, roundish red flowers on plants
from 6 to 12 inches high.

E. cyathoides, open purple flowers ; plant from
6 to 12 inches in height.

E. glandulosa and *lanuginosa*, A. H. vol. 3, ventricose yellow flowers on plants under a foot
high.

E. purea, tube-shaped white flowers ; plant rising
10 inches. B. C. 72.

E. cumulata, ocularia and *Solandra*, A. H. vol. 2,
roundish white flowers on plants from 6 to 12
inches high.

E. coccinea, c. major, Rollinsonia, Archeria and
erubescens, A. H. vol. 3, tube-shaped red and
flame-coloured flowers on plants from 12 to 18
inches high.

E. mucronata, globosa, princeps, B. C. 647, A. H.
vol. 2, *palustris*, B. C. 4, A. H. vol. 2, *pulchella*, B. C. 307, *scabriuscula*, B. C. 517, and
Sebana viridis, roundish red flowers on plants
from 12 to 18 inches high.

Erica rugata, a tube-shaped purple flower ; plant from 10 to 14 inches high.

E. tiaræflora, roundish purple flowers on a plant from 10 to 14 inches high.

E. Plukenetiana alba, tube-shaped white flowers ; plant from 12 to 18 inches high.

E. eriocephala, pill-shaped white flower ; plant from 12 to 18 inches high.

E. prægnans coccinea, carneola and *multiflora,* A. H. vol. 2, roundish red flowers on plants from 20 to 24 inches high.

E. vestita purpurea, tube-shaped flower ; plant from 20 to 24 inches high. B. C. 217.

E. Petiveriana, a tube-shaped yellow flower ; plant nearly 2 feet high.

E. glutinosa, roundish yellow flowers ; plant upwards of 20 inches high.

E. superba, viridiflora and *viridescens,* B. C. 233, roundish green flowers on plants from 20 to 24 inches in height.

E. planifolia and *setacea,* A. H. vol. 1, pill-shaped white flowers on plants rising 2 feet high.

E. vestita coccinea, tube-shaped red flowers ; plant upwards of 30 inches high. B. C. 55.

E. vestita fulgens, tube-shaped purple flowers ; plant above 3 feet high.

E. Leea, handsome yellow open flowers on a plant 3 feet high and upwards.

55

October.

Erica rosacea, open purple flowers ; plant from 6 to 12 inches high.

E. radiata, tube-shaped yellow flowers ; plant rising a foot in height. A. H. vol. 1.

E. horixontalis, tube-shaped red flowers ; plant from 12 to 18 inches high. A. H. vol. 2.

E. turgida, pyramidalis, B. C. 319, B. M. 366, and *perlata rubra,* roundish red flowers on plants from 12 to 18 inches high.

E. corrugata and *flava,* A. H. vol. 2, tube-shaped yellow flowers on plants rising 18 inches.

E. picta and *serratifolia,* A. H. vol. 1, roundish yellow flowers on plants from 12 to 18 inches high.

E. viridis, tube-shaped green flowers ; plant nearly 18 inches high. A. H. vol. 2.

E. denticulata alba, perlata, and *senectula,* roundish white flowers on plants from 12 to 18 inches in height.

E. cerinthoides, B. M. 220, *lanata,* and *c. capitata,* tube-shaped red flowers on plants about 2 feet high.

E. pulverulenta, roundish red flowers ; plant nearly 2 feet high.

E. filamentosa, tube-shaped purple flowers ; plant nearly 24 inches high. B. R. 6.

Erica purpurea, roundish purple flowers ; plant from 20 to 24 inches in height. B. C. 703, A. H. vol. 2.

E. aurea, tube-shaped yellow flowers on plants nearly 2 feet high. A. H. vol. 2.

E. pinastri, tube-shaped white flowers on an elegant pine-looking plant 2 feet high and upwards.

E. Eweria, E. pilosa, B. C. 606, and *E. speciosa,* A. H. vol. 2, B. C. 575, tube-shaped red and green flowers on plants nearly 2 feet high.

E. cerinthoides elata, open roundish red flowers on plants upwards of 2 feet and a half high.

November.

E. laxa, pill-shaped purple flowers ; plant from 12 to 18 inches high. A. H. vol. 3.

E. colorans, roundish purple flowers ; plant upwards of 20 inches high. B. R. 601, B. C. 224.

E. sulphurea, tube-shaped yellow flowers ; plant from 18 to 24 inches high. B. M. 1984.

December.

E. vestita rosea, tube-shaped red flowers ; plant upwards of 2 feet and a half high.

All the above species, and a number of others, are figured in Andrews's *Heathery,* and several of them in the B. M., B. C., and B. R. Those who live at a distance from the metropolis, may make a selection according to their own taste from the first work ; but those who can have an opportunity of seeing the plants at any of the great London cultivators, will be able to judge more correctly by inspection, especially as to the hardiness of the plant and its mode of growth. The flowers of heaths are in two grand divisions, the *tube-shaped* or long trumpet-like flowers, and the *roundish*, including the pill-shaped, cup-like and ventricose flowers. In making a selection, a due proportion of each should be fixed on ; a due proportion also of red, white, yellow, and purple, of both shapes ; and a due proportion of plants of the different heights from 6 to 30 inches.

The only soil in which heaths will grow is earth of peat : if any substitute can be found, it is in leaf-mould sifted very fine and mixed with fine sand. Earth of peat is obtained by collecting peats from bogs, or turf from the surface of peaty wastes and moist places, and laying the peats or turves in a heap to rot and moulder into earth. This they will require several years to do ; but in the mean while a portion of mould may be obtained whenever it is wanted, by turning the turves and sifting the fragments. Sometimes this peat is found without any mixture of sand ; at other times, where streams have run into the bog or lake while the peat was forming, it is mixed with fine sand that had been held in solution by the water. This last is the best sort of peat for the *Erica* family; and therefore where peat is not sandy naturally, fine white sand or sand of any colour, provided it be free from irony impregnations, should be procured and mixed with it. This sand admits the water to penetrate into the soil and reach the roots of the plants, and also to drain away from the roots so as not to rot them. Pots filled with pure peat earth are apt to be either hard, dry, and impenetrable to water, or otherwise as wet as a saturated sponge.

The climate for the heaths is not required to be warm during winter; if the frost is excluded, that will be enough. Some species, as the *E. persoluta* for example, will even bear to have the ground about their roots frozen without injury, provided it is not thawed in the sun, or too suddenly, or in a very warm temperature. In general the heaths may be kept in the coldest part of the green-house, and those not in flower in pits, well covered at night with mats, or prepared coverings of reeds or straw.

Heaths require a great deal of air and light, and therefore should be placed near the glass and near such glass as may be opened to admit

air every mild day in the year. They require
also very regular supplies of water; not much at
a time, but so frequently that the earth may never
get dry or the plant droop. Many kinds of plants,
if they have suffered for want of water, may be recovered by an abundant supply, and placing them
under a bell-glass on a little heat; but if once the
roots of a heath are thoroughly dried, no art of the
gardener will recover the plant. This is the true
reason why so many heaths are destroyed when
introduced as chamber plants, and also by gardeners who are ignorant of their nature.

Heaths are propagated by cuttings, seeds, and a
few by layers. In propagating by cuttings, the tender tops are taken at whatever season of the year
they begin to grow, which with most sorts is about
the month of June. Then take the extreme points
of the shoots, and with a sharp penknife cut off their
lower ends at right angles, placing the cutting on
the nail of the thumb, as in cutting the nib of a pen.
The cutting will be from three-quarters to an inch
long : strip off the leaves from the lower end to
nearly half the length of the cutting; and, in order
that this may be done without injuring the shoot,
use a sharp penknife or a pair of small scissars,
for the least bruise or wound spoils the cutting.
This done, dibble the cuttings into pots filled with
moistened white sand from pits, or with any small

sand from pits or rivers, or in default of that with powdered sandstone. When they are all planted, water the whole to fix them still better, and when the moisture has subsided, cover them with a small crystal or greenish crystal bell-glass fitted within the rim of the pot, and place them in the shade on a spent hot-bed, keeping them quite close till rooted. The free-striking sorts will have roots in two months, and the others at different periods from three to twelve months—most of them will be ready for transplanting into pots of the smallest size in the following March. Their rooting is easily known by their beginning to shoot, and then the bell should be taken off an hour or two daily.

Many Ericas ripen their seeds in this country, and of other sorts seeds are regularly obtained by the nurserymen from the collectors at the Cape of Good Hope. Imported seeds generally arrive in the winter, and should be sown early in the spring following, in pots filled with equal parts of peat and sand well incorporated; the seeds should be thinly covered with earth gently pressed down, and bell-glasses placed over them as over the cuttings. The soil must be kept moderately moist by gentle waterings, and in about six or seven weeks the seeds, if fresh, will begin to come up, when the glasses may be removed by degrees, and the pots kept near the glass and shaded from the mid-day sun till autumn, when they may be transplanted into pots of the smallest size.

Seeds which are saved in this country may be sown as soon as gathered, if they ripen before November; but if after that period, it will be better to preserve them till spring, and then treat them like foreign seeds.

Only a few heaths are propagated by layers, such as *E. Massoni, retorta petiolata,* and one or two other delicate sorts, which when layed require two years to throw out roots. On the continent most sorts of heaths are propagated by layers, because there they are ignorant of the easiest mode of managing cuttings.

For small collections, however, it will generally be found more economical and suitable to recruit the stock of heaths from the young plants of a nursery, than to attempt raising them by seeds or cuttings.

One of the best growers of heaths in Britain is a gardener of the name of Henderson, at Woodhall, in West Lothian. This judicious cultivator has had an extensive collection of *Ericæ* for upwards of thirty years under his care, and has given some account of his mode of management in a late volume (vol. iii. p. 323.) of the Caledonian Horticultural Society's Memoirs. He keeps his Ericas, he says, " at all times cool and airy, opening the glasses in winter when there is no frost, and letting the wind blow on them, and using no fire but in time of frost." " Never," he says, " shift any plant till the pot is quite full of roots. When the plants get large, several of them will continue in good health for three or four years without shifting, and flower well. I have plants of *E. retorta* here, in pots 7 inches in diameter, which are very bushy, being 18 inches across, and 14 inches high above the pot; *E. infundibuliformis,* two feet and a half in diameter, and 2 feet 9 inches high; *Erica pilosa,* between 5 and 6 feet high and 3 feet across, in pots 11 inches in diameter : these have not been shifted for five years, and are in high health, and covered with strong fine flowers from the mouth of the pot to the top of the plant." (Caled. Mem. iii. 327.)

" A prejudice," Page observes, " having spread that the culture of Heaths is difficult, one of the greatest ornaments of the green-house has hence of late been neglected ; although the method of culture is as easy and nearly as certain as that of the Geranium, but requiring a little more delicacy in the execution." (*Prodromus, &c.* art. *Erica.*)

One circumstance in favour of the culture of heaths is, that they are not subject to insects, or at least very rarely so.

Subsect. 3. GERANIUMS.

The *Geraniaceæ* come next in order. This is a
beautiful natural family of plants, comprehending
numerous species, herbaceous, suffruticose, and
shrubby or shrub-like, but all somewhat of a succulent or spongy nature. They have lately been
divided by botanists into several genera, of which
the chief are *Pelargonium, Erodium, Geranium,
Phymetanthus* (the *Pelargonium bicolor* and
tricolor of the nurseries), *Otidia, Hoarea, Jenkinsonia, Ciconium, &c.*, and they will probably
soon undergo other divisions; for so numerous
are the species become, by hybrids produced by
cultivators, that it will otherwise be difficult to
distinguish them. The Erodiums and Geraniums
are almost all hardy herbaceous plants; those which
are inhabitants of the green-house are chiefly
Pelargoniums, which differ from the Geraniums
in having the flower more like the bill of a
stork than of a crane. The genus *Pelargonium*
has lately been beautifully illustrated by Mr.
Robert Sweet in his *Geraniaceæ;* but as the
specific names there adopted are not yet in use
in all the nurseries, we shall first give a select
list of the names generally known, and then a
selection from the species published by Mr.
Sweet. These plants are in flower from February

to October or November; and by placing them in heat, some will flower during the whole of the winter. We shall enumerate them in the order of their flowering. The flowers of most of the sorts are so mixed in regard to colour, that it is almost impossible to class them in that respect; most of them are variegated with red, purple, scarlet and white. The height of all is from 1 to 3 feet.

February, March, and April.

Pelargonium dipetalum, a small flower but early; plant hardy and easily managed.

P. roseum, fine red flower, dark green leaves.

P. pictum, beautifully variegated flower, with a pink blotch, and leaves notched.

P. floribundum, showy flower, entire leaves, and very hardy plant.

P. laciniatum, small spotted flower, leaves much cut.

P. carneum, small flesh-coloured flower.

P. grenvillianum, splendid flower on a large-leaved plant.

P. pulchellum, bright red and white flower; leaves of a fine green.

P. ignescens and *ardens,* deep red flame-coloured flowers on plants with small roundish serrated leaves.

Pelargonium cordatum and *spurium*, very showy flowers on plants with heart-shaped leaves tall and very hardy.

May.

P. longiflorum and *longifolium*, the first with long petals, the second with long ovate leaves, both free flowerers and very hardy.

P. elegans, a beautiful and delicate flower; plant rather tender and easily injured by over-watering.

P. crenatum, odoriferous and hardy, flowers not very large.

P. Barringtonii, a splendid flower on a strong plant at one period considered the first of storksbills: it is a hybrid between *cucullatum* and another species.

P. cucullatum, a showy large-flowered plant, very hardy, with light green hooded hairy leaves.

P. speciosum, an elegant-flowered plant with smooth dark green leaves, and hardy.

P. grandiflorum, a large white flower; plant rather of slender growth.

P. variegatum, striped roundish leaves, fine scarlet flower.

P. crispum, small odoriferous flowers and small curled leaves.

F

Pelargonium gibbosum, rather difficult to keep;
the roots should have no water when the plants
are in a dormant state.

P. Beaufortiana, a very handsome species both as
regards the flower and the plant.

*P. alatum, althæoides, myrrhifolium, condupli-
catum, sidæfolium, acerifolium, quinatum, am-
plissimum, fuscatum, patulum, sororium, gratum, delphinifolium, cynosbatifolium,* and *spinosum,* species of different degrees of merit, but all
beautiful, tolerably hardy, and which are in perfection in May.

June.

P. auriculatum, a handsome-flowering plant with
green leaves covered with very minute hairs.

P. purpurascens, a fine dark flower, and hardy
plant.

P. zonale, one of the oldest plants of the family,
very hardy and of considerable beauty. Leaves
finely marked, and the plant, if trained on rods
or against a wall, will cover several yards and
reach 8 or 10 feet in height.

P. marginatum, an elegant-flowering species, but
rather tender.

P. pubescens, rugosum, and *rubens,* delicate flowering odoriferous plants.

Pelargonium papilionaceum, an old showy much
esteemed species.

P. glutinosum and *hispidum*, small-leaved delicate-
flowered plants.

P. glutinosum, dark green glutinous leaves and
foot-stalks ; flower small but of bright colours.
Plant much infested with the aphides, to be
destroyed most easily by smoking with tobacco.

*P. undulatum, virgineum, atrum, nervifolium,
pilosum, melananthon, ovale, glaucum, stenopeta-
lum, pubescens, crithmifolium,* and *ceratophyl-
lum,* are all elegant-flowering hardy plants
deserving a place where there is room.

July.

P. cuspidatum, an elegant-flowering hardy plant
with small smooth and slightly serrated leaves.

P. penicillatum, beautifully striped petals on a
plant not unlike *cuspidatum.*

P. betulinum, elegant flowers on a plant with
birch-like leaves on small woody twigs.

P. formosum, large showy and finely variegated
flowers on a hardy green-leaved plant.

P. capitatum, small red flowers on a plant highly
odoriferous.

P. ribifolium, elegant flowers on a plant with cur-
rant-like leaves and a rather soft spongy wood.

Pelargonium abrotanifolium, small bright flowers on plants with finely divided leaves.

P. tenuifolium, showy flowers and fine delicate leaves, but the plant hardy.

P. radiatum, large flowers on a hardy plant with lobed leaves.

P. lineare, punctatum, incrassatum, blattarium, grossularioides, lacerum, caucalifolium, diversiflorum, lateritium, and *australe,* free-flowering plants of different degrees of beauty, but all hardy and of easy culture.

August.

P. lobatum, handsome red flowers on leaves rough, dark green and deeply lobed.

P. triste, small flowers which smell delightfully during night.

P. odoratissimum, very small flowers on a dwarf plant with highly odoriferous leaves.

P. fragrans, a hardy sweet-smelling variety.

P. tricolor, a small plant with sea-green leaves and brilliant flowers, black, white, and red. *The Phymetanthus of modern authors,* see *Sweet's Geraniaceæ,* 96.

P. inquinans, an old variety with yellow-barked shoots and leaves, marked somewhat like *P. zonale.*

69

Pelargonium tetragonum, variegatum, and *angu-
losum,* old varieties of esteemed beauty and fragrance.

P. splendens and *fulgidum,* brilliant large flowers
and broad light green leaves; the plants hardy
and of easy culture.

*P. flavum, reniforme, heteroganum, crassicaule, pel-
tatum, lateripes, graveolens, Radula, denticulatum, semitrilobum,* and *alternans,* varieties and
species of different degrees of merit and all of
easy culture.

September.

P. bicolor, a brilliant flower much esteemed.
P. canariense, delicate-coloured flower, but hardy
plant.
P. scabrum, odoriferous flowers on rough-looking
leaves and stems.
P. gratum, elegant red and white flowers and
green leaves almost entire.
P. bipartitum, lemon-smelling flowers, and leaves
on small woody twigs.
P. exstipulatum, small flowers, and leaves somewhat resembling those of *P. odoratissimum.*
P. candidum, beautiful white flowers and large
light green leaves.
*P. balsameum, quinquevulnerum, consanguineum,
obtusifolium, Willdenowii, carnosum, lanceolatum,*

hybridum and *cortusæfolium*, species and varieties of different degrees of merit, all deserving
culture where there is room, and all hardy and
easily managed.

The above sorts are known by these names in
most nurseries at present; but as the new nomen-
clature of Sweet is coming fast into general use,
we subjoin a very select collection which flower
during the greater part of the summer, from the
" *Geraniaceæ*," with references to the figures in
that work.

Pelargonium striatum, Davey's Fairy Queen;
finely streaked flowers on roundish dark green
leaves. *S. G.* 1.

P. ignescens, fiery-flowered Storksbill; deep scarlet and black flowers on deeply divided dark
green leaves. *S. G.* 2.

P. blandum, blush-flowered Storksbill; light flower
on dark cordate leaves: it is sometimes called
the Waterloo and Diana Geraniums. *S. G.* 4.

P. obtusilobum, blunt-lobed Storksbill; dark red
flowers on leaves deeply 3-lobed. *S. G.* 8.

P. pannifolium, cloth-leaved Storksbill; blueish
light flowers on cordate leaves feeling like cloth.
S. G. 9.

P. Mostynæ, Mrs. Mostyn's Storksbill; red
flowers on pubescent reniform leaves: plant
very hardy. *S. G.* 10.

Pelargonium carduifolium, cockleshell-leaved Storksbill; large red flowers on rigid strongly-ribbed cucullate dark green leaves. S. G. 15.

P. multinerve, red flowers on kidney-shaped channelled leaves; plant very hardy, beginning to flower early in spring, and continuing in full bloom till late in autumn. *S. G.* 17.

P. adulterinum, Kutusoff's Geranium; deep red flowers from cordate three-lobed downy leaves; plant hardy, and one of the earliest-flowering species of the genus. S. G. 22.

P. solubile, dissolvible coloured Storksbill, or Duchess of Gloucester's Geranium; light red large flowers on kidney-shaped concave leaves, the plant very robust; water applied to the flower by the watering-pot or during a shower of rain dissolves the colour of the petals, a circumstance which happens to only a few of nearly the same coloured geraniums. S. G. 24.

P. eximium, select Storksbill, deep red flowers from large cordate undulate leaves; the plant a free grower and abundant flowerer. S. G. 26.

P. rubescens, Countess of Liverpool's Storksbill; large dark and light red flowers on cordate 5-lobed leaves; the plant tall, erect, and a free grower. S. G. 30.

P. Daveyanum, Davey's Storksbill; very dark red

and light red flowers, on a branchy plant with
cordate leaves. S. G. 32.

Pelargonium involucratum, large-bracted Storksbill; large striped flowers on a tall shrubby
stem, and kidney-shaped or cordate leaves.
There are several sub-varieties of this hybrid
known under different names, as " Commander
in Chief," " Davey's High Admiral," &c.: they
are all fine showy plants and free growers.
S. G. 33.

P. oblatum, oblate-leaved Storksbill; large red
flowers on very large dark green leaves (5 inches
by 6 inches), cordate and imbricate at the base,
the stem shrubby and growing to a great size,
very hardy. S. G. 35.

P. elegans, elegant Storksbill; finely striped, light-
coloured flowers on round leaves and a low
plant; it is rather delicate, and propagates best
by cuttings of the root. S. G. 36.

P. Seymouriæ, Mrs. Seymour's Storksbill; fine red
flowers with dark spots on a shrubby branching
plant with cordate leaves. S. G. 37.

P. pavonium, peacock-spotted Storksbill; deep red
spotted flowers on a shrubby branching plant,
with wedge-shaped cordate leaves; among the
handsomest hybrid Geraniums that have yet
been raised. S. G. 40.

73

Pelargonium floridum, abundant-flowering Storksbill. S. G. 41.

P. Lousadianum, Miss Lousadia's Storksbill.
S. G. 44.

P. Boyleæ, the Countess of Cork's Storksbill.
S. G. 50.

P. opulifolium, Guelder-rose-leaved Storksbill.
S. G. 53.

P. glaucum, glaucous-leaved Storksbill. S. G. 57.

P. sæpeflorens, frequent-flowering Storksbill.
S. G. 58.

P. bellulum, neat Storksbill. S. G. 60.

P. Breesianum, Breese's Storksbill. S. G. 64.

P. imbricatum, imbricate-petalled Storksbill.
S. G. 65.

P. cordatum, heart-leaved Storksbill. S. G. 67.

P. Thynneæ, the Marchioness of Bath's Storksbill.
S. G. 74.

P. calycinum, Brown's Countess of Roden. S.G. 81.

P. atrofuscum, dark brown-marked Storksbill.
S. G. 82.

P. macranthon, large-flowered Storksbill. S. G. 83.

P. Colvillii, Colvill's Storksbill. S. G. 86.

P. Baileyanum, Bailey's Storksbill. S. G. 87.

P. obscurum, darkened-petalled Storksbill. S.G.89.

P. Husseyanum, Lady Mary Hussey's Storksbill,
or Brown's Duke of York Geranium. S. G. 92.

P. bicolor, two-coloured Storksbill. S. G. 97.

Pelargonium villosum, villous Storksbill. S. G. 100.

P. difforme, Davey's Princess Augusta. S. G. 105.

P. Smithii, Smith's Storksbill. S. G. 110.

P. platypetalon, Oldenburgh Geranium. S. G. 116.

P. Scarbroviæ, Countess of Scarborough's Storksbill. S. G. 117.

P. formosum, variegated-flowered Storksbill. S. G. 120.

P. chrysanthemifolium, chrysanthemum-leaved Storksbill : one of the handsomest shrubby mules that have yet been raised. S. G. 124.

P. floccosum, Jenkinson's Rebecca. S. G. 129.

P. Watsoni, Watson's Storksbill. S. G. 130.

P. Youngii, Young's Storksbill. S. G. 131.

P. pulcherrimum, beautiful Storksbill. S. G. 134.

P. spectabile, showy Storksbill. S. G. 136.

P. Beaufortianum, Duchess of Beaufort's Storksbill. S. G. 138.

P. principissæ, Princess Charlotte's Storksbill : a beautiful hybrid nearly allied to *P. Beaufortianum*.

P. Brownii, Brown's Miss Rosa. S. G. 146.

P. Robinsoni, Robinson's Storksbill. S. G. 150.

P. Jenkinsoni, John Bull Geranium. S. G. 154.

P. æmulum, rival Storksbill. S. G. 160.

P. Murrayanum, Lord James Murray's Storksbill. S. G. 164.

Pelargonium venustum, comely Storksbill.   S. G. 167.

P. *Willsianum,* Mr. Wills's Storksbill.   S. G. 175.

P. *Broughtonianæ,* Lady Broughton's Storksbill. S. G. 181.

P. *lyrianthinum,* royal purple Storksbill, or More's Princess of Denmark.   S. G. 183.

P. *inscriptum,* marked-petaled Storksbill. S.G. 193.

P. *aurantiacum,* orange-coloured Storksbill. S. G. 198.

P. *calocephalon,* Tull's imperial Geranium, or pretty-headed Storksbill.   S. G. 201.

P. *incanescens,* whitish-leaved Storksbill.   S. G. 203.

P. *modestum,* modest Storksbill.   S. G. 204.

The above sorts are all hardy plants with very showy flowers.   They are of easy culture in the ordinary temperature and circumstances of a green-house,—a soil of loam with a little peat, and little water when not in flower. They may all be propagated by cuttings, cut across at a joint, and planted in the same sort of soil the parent plant grows in, with the addition of a little sand. They require neither a bell-glass nor to be placed on heat, but merely shading from the direct rays of the sun.

There are a number of other *Pelargonia* with

bulbous and tuberous roots; but we have not enumerated them because they are more curious than
showy, and less easy to manage than the others.
There are also a number of new genera, which Mr.
Sweet and other botanists have formed out of what
were formerly *Geraniums* or *Pelargoniums;* but as
these are chiefly curious plants, we have made no
selection from them here, as they will be found noticed in the general catalogue (Chap. IV.). It may
be proper to state that these genera are as follow,
viz. * *Campylia*, * *Hoarea*, *Monsonia*, * *Jenkinsonia*,
Phymetanthus, * *Otidia*, *Dimacria*, *Isopetalum*,
* *Ciconium* and *Grielum*, besides the old genera
Erodium and *Geranium*. Of these we would recommend *Phymetanthus* as being the old *Pelargonium tricolor* and *bicolor*, and one or two species
of the others which we have marked with an as-
terisk, provided there be room in the green-house.

The culture of all the *Pelargoniums* which we
have enumerated, is perhaps easier than that of
any green-house plant whatever. All that is necessary is to furnish them with regular supplies of
water, and to see that it drains away freely by the
hole in the bottom of the pot. The plants must
of course have a just share of light and air; and
as to light, indeed, they should be placed nearer
the glass than evergreen plants with coriaceous
leaves and woody shoots, such as the myrtle and

camellia. When they grow large, straggling, or unwieldy, they should be cut down or cut in; for the smaller the plants, in general, the larger and more vigorous the flowers and leaves. Seeds will ripen on most sorts, and should be carefully preserved, as nothing forms a more pleasing garden amusement for the ladies of a family, than saving these seeds, sowing them, and watching their progress in search of new sorts. The seeds may be sown in pots of light earth, or in a hot-bed, as soon as gathered; and when they come up and show a proper leaf, they may be transplanted into pots of the smallest size. Seedlings, from seeds saved and sown early in the season, will flower the succeeding autumn, and none will be later than a year in showing flowers; which is a great encouragement.

It is not an uncommon practice among lady amateurs to send their geraniums or a few of them when in flower to some of the nurserymen, to get them placed where they will be influenced by the genera of other sorts, and thus to try what new hybrid or mule will be the result. Nosegays of geranium blossoms are also procured, and suspended over plants in flower with a view to the same object.

A stock of camellias or heaths requires to be kept up by going to market; but no person need

ever be at a loss to keep up a stock of geraniums
by his own exertions. All the species which
have not bulbous or other thick roots are propagated by cuttings of the wood taken off at a joint
where it is beginning to ripen; and all the thick-
rooted kinds, by bits of root planted so as one
quarter or half an inch of the root may stand out
of the earth. Cuttings in this way may be taken
off at any time during the summer season ; and if
there is a cucumber frame at work, by plunging
the pots in it handsome plants may be obtained
in from one to three months from the time of taking
off the cutting. Many of the nurserymen plant
their cuttings in the open ground in a shady
border, and take them up and pot them in the
autumn. In some private gardens a number of
cuttings of the hardier kinds are every spring
planted in the open borders, where they make a
fine show during summer ; and though they are
killed by the first frost, it is easy planting more
cuttings the next season. In other gardens,
instead of planting cuttings in the borders, the
plants in pots when they grow old and stubby,
or any way unsuitable, or too bulky for the pot or
space limited for them in the green-house, are
turned out in the borders early in spring, and produce a farewell display of blossoms before they
are cut off by the autumnal frosts.

Subsect. 4. THE CITRUS TRIBE.

From the foregoing three genera we would recommend to be selected at least half the number
of pots and plants to be contained in any ordinary green-house. Next we would advise a
few of the citrus tribe, as beautiful evergreens,
most fragrant when in flower, and splendid when
in fruit. The following will form a very good
small assortment.

Citrus Aurantium, the common Orange, Maltese
variety; also the silver-striped; gold-striped;
myrtle-leaved.

C. nobilis, the clove or mandarin Orange, figured
in A. R. 608.

C. medica, the Citron; very large leaves and a fine
long yellow fruit.

C. medica, var. *Lenion*, the Lemon. The lemon
and the citron bear a considerable resemblance;
the fruit of the former is less knobbed at the
extremities, is rather longer and more irregular,
and the skin is thinner than in the citron; both
sorts are confounded by the dealers in the
London market.

The variegated-leaved lemon is a very handsome plant.

C. acida, the Lime; the leaves of this species are
almost quite entire and ovate; the fruit globular,
smooth, of a greenish yellow, shining, and the
flower very odorous.

Citrus decumana, the Shaddock; the leaves are ovate and long with winged footstalks, the wings almost as broad as the leaves; the fruit is spheroidal and grows to a large size, sometimes 7 or 8 inches in diameter; the plant is remarkably vigorous.

There are a great many varieties and sub-varieties of the above sorts; but those enumerated will form a very handsome collection, will look green at all seasons of the year, and fill the air of the green-house with a delightful odour when in blossom.

The culture of the orange and lemon tribe is very simple. They require to be planted in a loamy soil well enriched with decomposed dung, and the pots sufficiently drained; as they are apt to collect dirt on the leaves, they require to be frequently watered over the top in the evenings in mild weather, and even leaf by leaf washed with a sponge. They are also subject to the attacks of an insect called the orange bug, a sort of turtle-shaped scale of about an eighth of an inch in length or more, which may sometimes be seen on the leaves and small shoots. The best way to get rid of these vermin is to brush them off with a small painter's brush, and then wash the plant well with a sponge and common water : some use soap-suds and sulphur, but in the hands of ordinary practitioners water is safer and does just as well.

When the fruit begins to set, they ought to be thinned and only a few left on each plant. When this is neglected and the trees left to carry as many as they may be able to nourish, the consequence is a deficiency of shoots and blossom buds for the following year, and the tree left so weak by maturing so much fruit, that it ceases for a year or two to show either blossoms or fruit.

The orange is best propagated by grafting or budding on lemon or shaddock stocks; but as we cannot recommend this mode for renewing or keeping up a stock in a villa green-house, unless where a complete gardener is kept, we shall not enter into details. Orange trees endure for many years, even centuries, if well taken care of, so that they seldom require renewal. The Citrus tribe is also propagated by cuttings and layers; but it can seldom happen that either of these modes will require to be resorted to in a small green-house. However, as some may be disposed to amuse themselves in this way, we shall describe Henderson of Woodhall's mode of growing the orange from cuttings; and we believe no British gardener has ever been more successful than him.

Henderson has raised the orange and lemon

G

from cuttings for 40 years past, and considers it as by far the quickest mode of getting plants, either for bearing or grafting on. His directions are as follow: "Take the strongest young shoots, and also a quantity of two year old shoots, and cut both into lengths from 9 to 18 inches. Take the leaves of the lower part of each cutting to the extent of about 5 inches, allowing the leaves above that to remain untouched; then cut right across under an eye with a sharp knife, so as to leave a smooth unfractured section: when the cuttings are thus prepared, take a pot and fill it with sand, sort the cuttings so that the short ones may be all together, and those that are taller by themselves. Then with a small dibble plant the cuttings about 5 inches deep in the sand, and give them a good watering overhead to settle the sand about them; let them stand a day or two in a shady place, and then plunge the pots to the brim in a frame with bottom heat. Shade them well with a double mat till they have struck root; when rooted take the sand and cuttings out of the pot, and plant them into single pots in proper soil: plunge the pots again into a frame with bottom heat, and shade them with mats for 4 or 5 weeks, or till they are taken with the pots, when they may be gradually exposed to the light."—From various experiments, Henderson found that pieces of two years' old wood struck quite well; and in place therefore of putting in cuttings 6 or 8 inches long, he has taken off cuttings from 1 to 2 feet long and struck them with equal success. At first he put in cuttings only in the month of August; but now he puts them in at every season of the year, except when the plants are making young wood. With a gentle bottom heat, and close covering with hand-glasses or frames, they generally root in 7 weeks or 2 months. The citron he finds the easiest struck and freest grower; and he frequently strikes shoots of citron and lemon 18 inches long, and as soon as they are transplanted and taken with the soil in the small pots, he grafts other sorts on them. He strikes cuttings, engrafts, and buds the Citrus tribe at any season. (*Caled. Hort. Mem.* iii. 308).

Nothing can be easier than to raise young orange or lemon trees from the seeds found in those imported. Sow in pots and cover 2 inches, and they will soon come up with or without bottom heat. The succeeding spring transplant into small pots, and the end of that season or the third summer, send for a skilful gardener to bud on them whatever sorts may be desired. This is an easy, expeditious, and agreeable way for ladies and other amateurs to obtain young plants.

Subsect. 5. VARIOUS GENERA OF WOODY
GREEN-HOUSE PLANTS.

We shall now give the names of what we think
ought to form the remaining stock of a small or
even moderately large green-house, where the
object is, as we have before noticed, not botanical
curiosity, but verdant and elegant beauty. We
shall omit at present succulents and bulbs, in order
to treat of them in sections by themselves; and
partly because we think but very few of them are
admissible in the villa green-house. Though we
have placed the woody plants chiefly in the order of
their flowering, yet we have not adhered very
strictly to this rule,—sacrificing it occasionally, in
order to bring several species of a genus together,
thinking this more useful than scattering them
through different months.

January, February, and March.

Acacia verticillata, B. M. 110.
A. juniperina.
A. linifolia, A. R. 394.
A. stricta, A. R. 53.
A. longifolia, A. R. 204.
These are very hardy and most ornamental
plants. They are evergreens with long narrow

singular leaves, and are profusely covered with yellow flowers for 2 or 3 months together. They grow in sandy loam, with a little peat or leaf-mould intermixed, and are increased by cuttings taken off in the young wood, and planted in sand covered with a bell-glass and plunged in a gentle heat.

Salvia dentata, finely variegated flowers which come out in December and last till February. The plant grows in any soil and strikes readily by cuttings.

Daphne odora, purple flowers on a neat bushy evergreen shrub which grows in peat soil, and is propagated by grafting on the common Daphne of the woods.

Protea mellifera and *pendula*, most elegant evergreens with fine glistening silvery leaves; flowers in spikes, but not very remarkable. They grow in loam and peat, and are increased, but not very readily, by cuttings.

April.

Myrsine africana, an elegant myrtle-like evergreen, with shining leaves and small flowers in great abundance; very hardy; soil loam and peat: cuttings.

Cineraria amelloides, B. M. 249. Blue aster-like flowers which are very showy, and the colour, blue, is rare in green-house plants.

Cineraria aurita, purple-flowered, B. M. 1786.

C. cruenta, red-flowered, B. M. 406.

Acacia pubescens, B. M. 1263.

A. suaveolens, sweet-scented; beautiful evergreen shrubs. See Feb.

Myrtus, the Myrtle, 6 or 8 varieties, well-known evergreen shrubs, with a profusion of white odoriferous flowers.

Protea longiflora, Ex. B. 81.

P. speciosa. Knight's *Protea.*

Elegant evergreens.

Struthiola juniperina, B. M. 222. An elegant plant with needle-like leaves and small white flowers.

S. tomentosa, A. R. 334.

S. ciliata.

S. imbricata.

Elegant foliage and whitish flowers; the plants requiring a similar treatment to Heaths.

May.

Melaleuca tomentosa.

M. armillaris, A. R. 175.

Fringy tufts of fine deep red and myrtle-like leaves; plants hardy and easily managed.

Pæonia Moutan, A. R. 100.

P. M. var. Banksii, B. M. 1154.

Splendid rose-coloured flowers on hardy low deciduous shrubs, which will stand the winter in a cold frame; soil loam and peat, and propagation by dividing at the roots or cuttings.

Polygala mixta, B. M. 1714.

P. stipulacea, B. M. 1715.

P. alopecuroides, B. M. 1006.

P. speciosa, B. R. 150.

P. bracteolata, B. M. 345.

Purple and white pea-like flowers on elegant evergreen shrubs, flowering often for 2 or 3 months together; the soil peat and a little loam and sand, and care must be taken neither to overwater nor neglect watering; they are best propagated by seeds.

Daphne oleifolia, B. M. 1917. Blue flowers on a handsome low shrub that will stand the open air in sheltered situations, and which in a greenhouse flowers the greater part of the year.

Diosma purpurea, purple flowers on an odoriferous heath-like shrub, which grows in peat soil and propagates by cuttings.

Barosma serratifolia, B. M. 273.

B. latifolia, B. R. 38.

B. ovata, B. M. 1616.

Plants like Diosma; elegant evergreens, odori-

ferous, with clusters of white and red flowers:
soil loam and peat :—cuttings.

Lavatera maritima, a splendid display of pink
flowers, leaves blueish green, and the wood
spongy; the plant requires a large pot and a
sandy soil; it is increased chiefly by seeds.

Lotus jacobæus, B. M. 79. A fine dark purple
pea-flower, on a delicate winged-leaved plant
which blooms the greater part of the summer;
it grows in loam and peat, and is best increased
by seeds.

Salvia africana.

S. aurea, B. M. 182.

S. colorata.

Splendid scarlet flowers on sage-like plants, of
the easiest possible culture and propagation; they
will grow in any soil, and increase either by seeds
or cuttings.

Genista canariensis, B. R. 217. Profusion of yellow flowers on an elegant little shrub which
grows freely in sandy loam, and is increased by
cuttings or seeds.

Gnidia pinifolia, B. R. 19.

G. simplex, B. M. 812.

G. oppositifolia, B. R. 2.

G. sericea, A. R. 225.

Twiggy evergreens with small leaves finely
mixed with yellowish whitish flowers; soil peat

and loam :—increase by cuttings in sand under a bell-glass in the manner of Heaths.

Cistus vaginatus, B. R. 225.

C. ladaniferus, B. M. 112.

C. incanus, B. M. 43.

Large flowers of momentary duration, generally whitish with yellow or dark purple spots ; the plants very hardy and increased by cuttings ; soil peat and loam, with sand.

Helianthemum formosum, B. M. 264.

H. halimifolium.

H. algarvense, B. M. 627.

H. lævipes, B. M. 1782.

Showy fugacious flowers like those of Cistus ; the plants hardy and of easy culture in sandy loam, or loam and a little peat.

Malva capensis. Small red flowers on an upright plant which is in bloom most of the summer. It is readily grown in any soil, and increased by seeds which it produces in abundance.

Correa and *Pittosporum* are hardy evergreen genera that stand the open air in Jersey and Guernsey ; they have white flowers which appear in this month ; but, excepting to fill up a very cold part of the green-house, or for the sake of their fine foliage and bushy shapes, few besides *C. speciosa* can be recommended in very select collections.

Pultenæa daphnoides, B. M. 1394.

P. obcordata, A. R. 574.

P. stricta, B. M. 1588.

Elegant orange-coloured pea-flowers, on singular Australasian evergreen shrubs, which grow in loam and peat, and are increased by cuttings and seeds.

June.

Crotalaria purpurea, B. R. 128.

C. pulchella, B. M. 1699. Large purple pea-flowers on handsome compound leaves : the plant deciduous, and of easy culture and propagation.

Epacris purpurascens, B. M. 844.

E. pulchella, B. M. 1170.

E. grandiflora, B. M. 982. Fine scarlet and crimson coloured flowers on plants of easy culture in peat and loam, and increased by cuttings and seeds.

Illicium floridanum, B. M. 439.

I. parviflorum. The first a fine large magnolia-like red flower, and the other a large yellow flower on broad-leaved evergreens, shining reddish and odoriferous. The plants easily preserved in sandy loam or loam and peat; propagation by cuttings but tedious.

Convolvulus canariensis, B. M. 1228.

C. cneorum, B. M. 459.

C. linearis, B. M. 289. Elegant purple and white flowered twiners with shrubby stems which answer to be grown in pots to twist round three or more rods or wires.

Indigofera psoraloides, B. M. 476.

I. cytisoides, B. M. 742. Delicate purple flowers on slender plants with light green glaucous leaves; the plants grow in sandy soil in very small pots, require but little water, and ripen seeds freely.

Gnaphalium eximium, A. R. 654.

G. grandiflorum, A. R. 489.

G. fruticans, B. M. 1802.

G. congestum, B. M. 243.

G. patulum. Yellow heads of chaffy everlasting flowers on white downy-leaved plants; the flowers odoriferous and chiefly valued on account of their duration as well on the plant as when gathered; soil peat and sand; propagation by cuttings.

Brunia nodiflora.

B. alopecuroides. Elegant shrubs with fir or heath-like leaves thickly covering their shoots and abundance of yellow flowers: culture as in *Erica.*

July.

Beaufortia decussata, B. R. 13. Splendid red
flowers on a myrtle-leaved plant; hardy, grown
in loam and peat, and propagated by cuttings.

Swainsonia galegifolia, B. M. 792.

S. coronillifolia, B. M. 1725.

Red and purple pea flowers on pinnate leaves;
the plants hardy and of easy culture.

Sutherlandia frutescens, B. M. 181. Profusion of
scarlet pea flowers and fine glaucous compound
leaves; the plant hardy and readily increased
by seeds, cuttings, or pieces of the root.

Coronilla glauca, B. M. 13.

C. valentina, B. M. 185.

Profusion of fine yellow flowers on elegant little
trees with glaucous pinnate leaves which grow in
any soil and seed freely.

Fuchsia coccinea, B. M. 97. A well-known favour-
ite, and one of the most elegant inhabitants of
the green-house. The whole plant has a purplish tinge, and the drooping flowers are scarlet and purple. It grows in leaf-mould, and
requires little water in the winter season when
dormant; cuttings root readily in sandy loam,
leaf-mould, or peat.

Hermannia flammea, B. M. 1349. Abundance
of scarlet and red flowers on a rugose-leaved
branchy shrub of the easiest culture and propagation.

Lavendula pinnata, B. M. 400.

L. dentata, B. M. 401.

Fine scented blue flowers on lavender-like plants which grow freely in sandy soil, and are increased by cuttings or seeds.

Cassia corymbosa, B. M. 633. Deep yellow pea
flowers on glaucous pinnate leaves; plant hardy
and of easy culture and propagation in light
soil.

Jasminum odoratissimum, B. M. 285.

J. grandiflorum, B. M. 91.

J. azoricum, B. R. 89.

Most odoriferous white and yellow flowers on pinnate leaves; the plants requiring support by training on a wall or trellis, or by props, but in other respects of easy culture, and propagated by cuttings or layers.

Heliotropium peruvianum, B. M. 141.

H. corymbosum, B. M. 1609.

Purple and white flowers in bunches smelling like new made hay; the plants low shrubs, deciduous, requiring little water when not in a growing state, and the warmest part of the green-house when in flower; they are readily increased by cuttings.

Buddlea salvifolia. Round heads and bright yellow flowers on a deciduous shrub with long
narrow rugose sage-like leaves; soil, loam and
peat, and propagation by layers.

Platylobium formosum, B. M. 469. Large orange
pea flowers on an elegant pinnate-leaved shrub
of easy culture and propagation.

August.

Nerium Oleander, varieties *alba* and *splendens*.
N. odorum, varieties flesh-coloured and double-
flowered.

Splendid red and white flowers, which when
the plant is large and has plenty of room make a
magnificent show for two or three months together;
the best soil is loam with a little sand and peat;
the plants require to be abundantly watered when
in flower, and the leaves, which are apt to contract
dust, should be occasionally sponged over to clean
and refresh them. These plants grow eight or
ten feet high where they have room, are very
hardy, and are increased by cuttings planted in
sand under a bell.

Chironia Jasminoides, B. R. 197.
C. linioides, B. M. 511.
C. baccifera, B. M. 233.
C. frutescens, B. M. 37.

Showy lake-coloured flowers on elegant little
spongy-wooded light green shrubs, which grow
in peat soil, and are rather difficult to strike by
cuttings.

Eutaxia myrtifolia, B. M. 1274. Deep orange

pea flowers on an elegant ever-green shrub with myrtle-like leaves and of easy culture in loam and peat, propagated by cuttings.

September to December.

Phlomis Lychnites, B. M. 999.   A sage-leaved whitish rugose plant with singular flame-coloured flowers.   It grows best in sandy loam or loam and lime rubbish kept dry.
Celsia Arcturus.
C. Cretica, B. M. 964.
C. lanceolata.

Biennial plants with brilliant scarlet, red, and whitish flowers which come out in abundance for three or four months together.   Any soil suits these plants; and they will live if protected by a cold frame, though in that situation they do not flower freely.   They are readily increased by cuttings or seeds.

Gordonia Lasianthus, B. M. 668.   Fine large yellow flowers of the character of *Lavatera,* on a smooth-leaved plant of easy culture in loam and peat, and increased by cuttings.
Crotalaria purpurea, B. R. 128.
C. pulchella, B. M. 1699.

Beautiful purple pea flowers on pinnate glaucous-leaved plants, hardy and of easy culture in sandy loam.   Propagation by seeds or cuttings.

Spielmannia africana, B. M. 1899. Fine scarlet flowers which continue coming out all summer and often to the middle of December. The plant is of the easiest culture and propagation.

Aloysia citriodora, B. M. 367. (formerly *Verbena triphylla*). Purplish white flowers in elegant spikes; but the plant is chiefly cultivated for the delightful odour of the leaves when bruised. It grows well in rubbish and loam, andis as easily increased by cuttings as the willow.

Clematis brachiata, B. R. 97.

C. aristata, B. R. 238.

C. calycina, B. M. 959.

Whitish flowers on branchy shrubs which require the support of other plants or of props. The flowers of no brilliance, but valued as coming out in the two last months of the year.

Statice pectinata.

S. sinuata, B. M. 71.

S. mucronata and *echioides.*

The first a shrub, the next two herbaceous perennials, and the last a biennial, all with showy red and purple flowers and evergreen glaucous leaves. They grow in sandy soil and are easily propagated by cuttings, division at the root, and sometimes by seed.

Linum trigynum, B. M. 1100. Fine yellow flowers which come out in November and last till February; the plant not difficult to preserve in

peat soil, and increased, though slowly, by
cuttings.

Westringia rosmariniformis, A. R. 214.   A rose-
mary-looking shrub with white flowers, of
easy culture in sandy loam, and increased by
cuttings.

Trachelium diffusum. Fine blue-bell-shaped flowers on a plant of no great beauty as to foliage,
but of the easiest culture and multiplication.

Arctotis acaulis, B. R. 122.

A. tricolor, B. R. 131.

A. maculata, B. R. 130.

A. aspera, B. R. 34.

A. aureola, B. R. 32.

Splendid orange, white, and purple flowers on
plants of no great beauty as to foliage, but very
hardy; and of the easiest culture and propagation
in any soil.

Dracocephalum canariense, the Balm of Gilead.
A well-known hardy plant, powerfully fragrant,
of the easiest culture in any soil, and increased
by seeds which it produces in abundance.

Gardenia florida, single and double.   This plant,
though properly belonging to the stove, may
yet be cultivated in gardens where there are
hot-beds and a green-house.   It is so highly
odoriferous that it is very desirable to have in a
collection, more especially as by judicious treat-

ment it can be made to flower in December, January, and February. Being a low shrub, it may be kept in a flued pit or in a cucumber frame, in a moist heat and near the glass, till it comes into flower, when it may be removed to the green-house or chamber stage. The great want of this plant is abundance of heat and light; it prefers a rich loamy soil well drained, and should be kept very clean by washing with the sponge. It is propagated by cuttings, but not very readily.

Lotus jacobæus, B. M. 79.

L. hirsutus, B. M. 336.

Purple pea-flowers, on glaucous pinnate leaves; the plants continuing in bloom almost all the year. They grow in sandy loam and are readily increased by seeds.

Rosa Banksiæ, B. M. 1954.

R. sinica.

R. semperflorens, B. M. 284.

R. odorata.

Dwarf roses which it is desirable to have in a green-house; the first is a fine ever-green and very hardy, and the others are esteemed for their abundance of flowers.

Elichrysum speciosissimum, A. R. 51.

E. fulgidum, B. M. 414.

E. proliferum, B. R. 21.

Elichrysum canescens, B. M. 420.

E. sesamoides, B. M. 425.

E. fasciculatum, A. R. 242.

E. fragrans, A. R. 100.

Yellow paleaceous flowers of long duration, commonly called everlasting flowers, which grow on whitish tomentose-leaved low plants of easy culture in peat soil, and increased by cuttings.

We will not say that the above enumeration contains all the most select woody plants which are inhabitants of the green-house ; but we will affirm that there is not a single species named therein that is not highly beautiful, and well deserving the appellation of select. These, with the Camellias, Geraniums, Ericas, and Citron tribe before enumerated, will, if assembled together without the intermixture of merely curious botanists' plants, and to the exclusion of all bad or sickly specimens, produce an effect beyond any thing the imagination can well picture out—at all events, an effect such as we meet with in hardly any green-house at present, on account of the random mode of bringing plants together. It is thought quite enough if plants which require the climate of the green-house are obtained, no matter what may be their appearance, the colour of their flowers, time of flowering, or whether they are deciduous or evergreen. Every thing is obtained, it is thought, if a great assortment of species are collected, though not one of them should have half the beauty of a common myrtle. We venture to affirm, and we request it may be taken into consideration by all admirers of plants not scientific botanists, that if there is such a thing as fine foliage, showy flowers, brilliant colours and elegant shapes, then three-fourths of the plants which require to be grown in green-houses have no claim to these appellations. On the contrary, we affirm that three-fourths are plants of meagre foliage, obscure dingy flowers, and uncouth straggling shapes. Let us leave these plants, then, to the botanist, who views them very properly with interest as component parts of his system, as links in the great chain of vegetable being.

In making the above selection we have had five leading objects in view :

Firstly, To ensure perpetual verdure in the green-house ; for which purpose we have chosen chiefly evergreens, and admitted only such deciduous plants (as *Fuchsia, Aloysia,* &c.) as are remarkable for the beauty of their flowers or their odour.

Secondly, To ensure a perpetual display of blossoms : for this end we have admitted none but what are considered as free flowerers ; that is, such as, with the usual culture, send out a profusion of flowers at their blooming season.

Thirdly, To ensure not only abundance of blossoms every month, but a due proportion of blossoms of each of the usual colours. It is a great drawback to the beauty of collections of plants, both exotic and indigenous, that the yellow colour is so predominant in the flower. Reds, scarlets and whites are the colours that contrast best with the green hue of foliage, and we have been most particular in selecting an equal number of species of these colours as of the yellows. It is a remarkable fact, that there are very few green-house plants that have blue flowers ; and for this reason we have been particularly attentive to get the proper assortment of them. *Cineraria, Diosma, Daphne, Indigofera, Statice, Convolvulus, Lavandula, Podalyria, Trachelium,* but especially *Polygala,* are valuable genera as producing blue or purple flowers ; though some of the species, as of *Statice* and *Trachelium,* are not quite so hand some in foliage as could be wished. Purple is the next colour in scarcity to blue ; but of this there is a fine resource in the genera *Pelargonium* and *Erica.*

Fourthly, To ensure a fragrant smell in the green-house every day of the year. This will be found accomplished by the *Acacia* family in January, February, and March ; by Myrtles and

Geraniums in April and May; by the *Citrus* tribe in June; and by an abundant variety during the rest of the season till September, when the fragrance of the green-house atmosphere must be kept up by tree mignionette, a plant of which there ought to be two or three specimens in the house throughout the year. (*See* Sect. 4. *infra.*)

Fifthly, We have chosen plants of easy culture, and which flower well when of a small or moderate size.

Subsect. 6. *Succulent Green-house Plants.*

The succulent plants proper to be admitted into a small green-house are in our opinion very few. Succulents cannot in general be considered as beautiful; they are curious and some of them oddities, and of forms which surprise at first sight; but who takes that pleasure in contemplating the leafless *Stapelia* or the grotesque *Cactus*, (however extraordinary the flowers of some species may be,) which he does in looking on a Rose or a *Camellia*? None whose tastes are not vitiated or singular, or who do not look solely with the eye of science. One or two curious or ugly objects, however, may be admissible to show that there are such things; and of these we shall select a few of the hardiest, and of those which have the finest flowers, arranging them according to their time of flowering.

January.

Mesembryanthemum luteo-viride. Yellowish-green flowers on a low spreading plant.

M. perviride. Dark green flowers on a herbaceous thick-leaved tuft rising only an inch or two from the pot.

February and March.

Mesembryanthemum semicylindricum.

M. gibbosum.

M. aureum, B. M. 262. Fine golden flowers on a bushy straggling plant which often continues in bloom for several months together.

M. maximum. Moonshaped leaves!

April and May.

Mesembryanthemum pulchellum.

M. coccineum, B. M. 262. Scarlet flowers.

M. micans, B. M. 448. Glittering.

June.

Aloe acuminata, or Hedgehog Aloe, B. M. 757.

A. plicatilis, Fan, B. M. 457.

A. variegata, partridge-breast, B. M. 513.

A. obliqua, broad-marbled, B. M. 979.

Sempervivum tortuosum, B. M. 296.
Mesembryanthemum spectabile, B. M. 396.
M. stelligerum, B. M. 70.
M. aurantium.
M. capitatum.

July.

Cotyledon ovata, B. M. 321.
Agave americana, A. R. 438.
Aloe picta, B. M. 1323.
Aizoon glinoides.
Mesembryanthemum inclaudens, A. R. 388.

August.

Cacalia Kleinia.
C. Ficoides.
Mesembryanthemum conspicuum.
M. expansum.
M. bellidiflorum.
M. canaliculatum.

September to December.

Mesembryanthemum tigrinum.
M. depressum, B. M. 1866.
M. nitidum, B. M. 326.
M. confertum.
Sempervivum arboreum, B. R. 99.

Anthericum revolutum, B. M. 1044.

Septus capensis, A. R. 90.

Anacampseros arachnoides, B. M. 1368. Cobweb
   Anacampseros.

A number of other succulent plants might be
enumerated, and especially if the genera *Cactus*
and *Stapelia* were admitted, as these, though properly dry-stove plants, will live in a green-house :
but the above specimens are much more than
enough for any small or showy green-house; and
we must confess we would rather none were admitted, than see one half of them introduced and
occupying the place of much finer plants.

Green-house succulents are of the easiest possible culture and propagation ; lime rubbish and
loam, or sand and loam, is the best soil ; the pots
being well drained, and care being taken never
to over-water them. Succulents with very thick
leaves, as the *Aloe, Crassula,&c.*, will live for months
without any water, and Mesembryanthemums require it chiefly when they are in flower. None
of the succulents, excepting the last-mentioned
genus, should be set out in the open air during
summer, as they are very apt to be rotted by rain
and cold. Cuttings allowed to dry a day or two
after they are taken off, and then planted in sandy
loam, root speedily and certainly; and the *Mesem-*

bryanthemum may be propagated in this way to any extent, and so as to have abundance of plants to plant out in the open borders of the flower-garden, to flower during August and September if the season is dry and warm. The *Mesembry-anthemum* indeed is the only genus of green-house succulents which has any pretensions to beauty in the flower : some admire them on account of their flowers and extreme hardiness; and one gardener has carried their culture to a considerable extent in pits ; in these he plants them without pots, and chiefly the species *inclaudens, aurantium, perfoliatum, deltoides,* and *barbatum.* The strong growing kinds are put towards the back, and the dwarf ones in the front of the pit. They grow vigorously, he says, and " flower in a superior manner to what they do in small pots : nothing can surpass the brilliancy of their blossoms in a bright summer day, and many of them continue flowering during the winter. All the culture they require is thinning, and protection by mats over the glass in severe weather. In summer the sashes may be taken off, and the soil covered with black stones like rock-work to attract and refract the heat." (*Hort. Trans.* v. 274.) There is no flue to this pit, but it is protected by being placed against the front wall of a hot-house.

Subsect. 7. *Bulbous-rooted Green-house Plants.*

There are a considerable number of bulbous-rooted plants which are inhabitants of the green-house, and a number of them usually grown in the dry-stove will live there. This class of plants however have no show excepting when in flower, and half the year at least they have not even leaves, being wholly dormant under the soil. We advise but very few of them, therefore, to be admitted into the villa green-house, and these few we think should be brought to it only a few days before they come into flower, and taken from it immediately after their blossoms begin to fade, and kept in frames or pits all the rest of the year. We shall enumerate some of the most showy green-house bulbous plants, chiefly in the order of their flowering.

March to May.

Cyclamen persicum, B. M. 44. Purple violet-like flowers on coriaceous roundish leaves, which lie close to the ground.

Gladiolus Watsonius, B. M. 450 and 569.

G. quadrangularis, B. M. 567.

G. viperatus, B. M. 688.

G. hirsutus, B. M. 574.

G. hastatus, B. M. 1564.

Gladiolus gracilis, B. M. 562.

G. cuspidatus, B. M. 582.

G. undulatus, B. M. 538.

G. Milleri, B. M. 632.

Trichonema cruciatum, B. M. 575.

T. speciosum, B. M. 1476.

Geissorhiza excisa, B. M. 584.

Hesperantha radiata, B. M. 573.

H. pilosa, B. M. 1475.

H. falcata, B. M. 566.

H. cinnamomea, B. M. 1054.

Sparaxis tricolor, B. M. 381.

S. t. sanguineo-purpurea, B. M. 1482.

S. t. violaceo-purpurea, B. M. 1482. f. 2.

S. t. roseo-alba, B. M. 1482. f. 3.

S. bicolor, B. M. 548.

S. grandiflora, B. M. 541.

S. g. striata, B. M. 779.

S. g. Liliago, B. R. 252.

Ixia linearis, B. M. 570.

I. capillaris, B. M. 617.

I. aulica, B. R. 1018.

I. patens, B. M. 522.

I. leucantha.

I. hybrida, B. M. 128.

I. conica, B. M. 539.

I. monadelpha, B. M. 607.

I. flexuosa, B. M. 624.

Ixia curta, B. M. 1378.

I. columellaris, B. M. 630.

I. retusa, B. M. 629. The flowers are odoriferous, and appear in January and February.

I. scillaris, squill-like flowers, also in January and February.

Tritonia crispa, B. M. 678.

T. squalida, B. M. 581.

Watsonia spicata, B. M. 523.

Lachenalia contaminata, B. M. 1401.

L. lucida, B. M. 1372.

L. racemosa, B. M. 1517.

L. pustulata, B. M. 817.

L. purpureo-cærulea, B. M. 745.

L. bifolia, B. M. 1611.

L. rosea, A. R. 296.

L. unifolia, B. M. 766.

L. tricolor, B. M. 82.

L. quadricolor, B. M. 588 and 1097.

L. pendula, B. M. 590.

June.

Amaryllis formosissima, B. M. 47.

A. blanda, B. M. 1450.

A. vittata, B. M. 129.

A. purpurea, B. M. 1430.

Amaryllis coranica, B. R. 139.

A. aurea, B. M. 409.

A. curvifolia, B. M. 725.

A. corusca, B. M. 1089.

A. Johnsoni, the *A. spectabilis* of B. C.

A. venusta, B. M. 1090.

A. radiata, A. R. 95.

A. undulata, B. M. 369.

A. humilis, B. M. 726.

A. flexuosa, B. R. 172.

A. longifolia, B. M. 661.

A. revoluta, B. M. 915 and 1178.

Pancratium canariense, B. R. 174.

Agapanthus umbellatus, B. M. 500.

Antholyza vittigera, B. M. 1172.

A. æthiopica, B. M. 561.

Gladiolus versicolor, B. M. α. 1042.

G. tristis, B. M. 272.

G. carneus, B. M. 591.

G. augustus, B. M. 602.

Babiana tubiflora, B. M. 847.

B. tubata, B. M. 680.

B. spathacea, B. M. 638.

Aristea melaleuca, B. M. 1277.

Lapeyrousia corymbosa, B. M. 595.

Witsenia corymbosa, B. M. 895.

Lachenalia angustifolia, B. M. 735.

L. unicolor, B. M. 1373.

Lachenalia nervosa, B. M. 1497.
L. sessiliflora, A. R. 460.

July.

Hæmanthus coccineus, B. M. 1075.
H. rotundifolius, B. M. 1618.
Strumaria gemmata, B. M. 1620.
Crinum pedunculatum, B. R. 52.
Cyrtanthus collinus, B. R. 167.
C. obliquus, B. M. 1183.
C. uniflorus, B. R. 168.
Brunsvigia Josephinæ, B. R. 192 and 193.
B. multiflora, B. M. 1619.
Amaryllis equestris, B. M. 305.
A. e. major, B. R. 234.
A. corusca, B. M. 1089.
Trichonema pudicum, B. M. 1244.
T. roseum, B. M. 1225.
Hesperantha graminifolia, B. M. 1254.
Ixia columellaris, B. M. 630.
Tritonia capensis, B. M. 1531.
T. Rochensis, B. M. 1503.
T. miniata, B. M. 609.
Watsonia roseo-alba, B. M. 537.
W. rosea, B. M. 1072.
Gladiolus cardinalis, B. M. 135.
Babiana purpurea, B. M. 1052.

Babiana villosa, B. M. 583.

Morea Pavonia, B. M. 1247, and various other species not exactly bulbous-rooted.

Lachenalia isopetala.

August.

Amaryllis aurea, B. M. 409.

A. sarniensis, B. M. 294. The Guernsey Lily, a well-known bulb, annually imported from Guernsey.

A. flexuosa, B. M. 172.

A. revoluta, B. M. 915 and 1178.

Hæmanthus rotundifolius, B. M. 1618.

H. quadrivalvis, B. M. 1523.

Strumaria gemmata, B. M. 1620.

Crinum pedunculatum, B. M. 52.

Tritonia capensis, B. M. 1531.

T. uvaria, B. M. 758.

Hesperantha graminifolia, B. M. 1254.

Lapeyrousia fissifolia, B. M. 1246.

Iris moræoides, B. M. 693.

Polianthos tuberosa, B. M. 63. The Tuberose, a most odoriferous plant, of which there are double and single varieties. The roots are annually imported from South America and Italy.

Ferraria undulata, B. M. 144.

Ferraria antherosa, B. M. 751.

Tigridia pavonia, B. M. 532.
Lachenalia rubida, B. M. 993.

September to December.

Watsonia marginata minor, B. M. 1530.
Gladiolus brevifolius, B. M. 727 *and* 992.   Flowers in December and January.
Witsesia maura, A. R. 5. Flowers in November, December and January.
Hæmanthus tigrinus, B. M. 1705.
H. quadrivalvis, B. M. 1528.
H. lanceæfolius.
Strumaria filifolia.
S. stellaris.
Amaryllis pumila.
Tritoma media, B. M. 744.
T. pumila, B. M. 764.
Ornithogalum niveum, B. R. 235.
O. lacteum, B. M. 1134.
O. revolutum, B. M. 653.
O. elatum, B. M. 528.
Scilla brevifolia, B. M. 1468. Flowers in December and January.

The above list comprises most of the choice green-house bulbs at present in cultivation : it contains ten times the number which we would wish to see introduced in the villa green-house (for bulbs assort as ill with woody evergreens, as do the succulent tribe) : but as all of them may be better grown and flowered in a flued pit, than any where else, we have made our catalogue more ample for the benefit of such as may resort to this plan.

The culture of bulbs is sufficiently simple, provided two points be strictly attended to : the first is, to take care not to injure their leaves, and to keep them near the light; and the second is, when the plants have done growing, to give them little or no water. From ignorance of the importance of attending to these two points, it is not uncommon to see bulbs in green-houses, which have lived for several years, and never shown any blossoms, especially of the genera *Crinum* and *Amaryllis*. Whether a bulb will blossom or not, depends entirely on its culture during the foregoing season ; that is, on whether it was so circumstanced as to bring its leaves to perfection. If the leaves were fully grown and properly exposed to the influence of the light, then the sap will have been duly elaborated by them, and an embryo flower formed in the bulb; if otherwise, no embryo will have been formed, and no culture whatever during the succeeding year will effect the production of a flower during that year. All bulbs have a certain period of the year, in which

they are in a dormant state. This in a state of
nature, is invariably after the seeds are ripened;
but as in a green-house many or most of this family do not ripen seeds, the gardener is required
to watch the period when the leaves show indications of decay, and then to lessen supplies of
water, and shortly afterwards to cease from watering
altogether, till the season returns, when the bulbs
regerminate.   The bulbs during this period are,
on the whole, best kept in the pots, under the soil in
a dry shady place, and in the same temperature
as that in which they are in the habit of growing.
Some bulbs may be taken out of the soil and kept
some time in papers; but if this is done for more
than a week or two, it tends to weaken the bulb.
The greater part of exotic bulbs should be taken out
of the pot, and repotted in fresh soil, a week or
two before their period of regerminating.

Loam with a little sand, vegetable mould, or
mould of spit dung, forms a compost or soil in which
almost all bulbs will thrive.

The *Amaryllis* requires a richer loam than most
bulbs, and *Ixia* and *Gladiolus* a soil rather more
sandy than the general average : equal parts of
peat and sand answer well for most of the Cape
or African bulbs, and loamy soil for those of the
East Indies.

All bulbs require to be placed near the light, and

they should have abundance of air in mild weather, and plenty of water whilst in a growing state.

Bulbs are mostly propagated by off-sets; and some, as the *Ixiæ* and *Gladioli*, afford seeds. Tunicate bulbs, as some of the Ornithogalums, if cut over a little above the middle and the root end planted with the section exposed, will form numerous little gems near the margin of the outer coat; and of most scaly bulbs, if a leaf in full growth be stripped off close down to the base of the bulb, and planted in light sandy soil, it will produce young gems at its base, and propagate by that means.

Sect. IV. *Hardy and half-hardy Plants that may be admitted in the Green-house.*

The next subject of consideration is the hardy and half-hardy plants that may deserve admission to a certain extent in the green-house; and these we shall, as before, notice in monthly order.

January, February, and March.

Early in the beginning of the year, or even in the preceding November and December, some bulbs should be planted in pots of sandy loam, plunged in the open garden, and covered with rotten tan, leaves, or litter. These after they have rooted may be taken up and placed for forcing in a pit or hot-bed, if such are in use; or kept some time in the kitchen or laundry, even though in the dark, till they protrude their leaves an inch or more, and then they may be set in the green-house. Some plant in pots, and set at once in the green-house, and others plant and set at once on heat, or in the kitchen, or other warm room; but it is a great advantage to bulbs to be plunged in cold earth for some time after planting, as it causes them to send out more numerous and vigorous roots, and when once this is done they may be placed on heat or in the green-house. Even when bulbs are to be forced in a regular hot-house, and also when they are to be blown in glasses of water, it is of the greatest advantage to root them in cold earth in the first place; for put on heat or on water, the roots never protrude freely, and the leaves and flowers are in consequence imperfectly nourished. The sorts of bulbs which may be admitted in the green-house are the following, and those imported are generally the best blowers.

Polyanthus Narcissus, several varieties; say six pots.

Jonquil Narcissus; two or three pots.

Hyacinths; a dozen pots.

Persian Iris for its odour; one plant will perfume a whole green-house.

Chalcedonian and Snakeshead Iris; a pot of each.

Duc van Thol Tulip, Snowdrop, Aconite Crocus, *Dens canis* and *Scilla verna*; not more than one small pot of each for the sake of variety.

Some persons are particularly attached to the culture of bulbs, in pots or water glasses. To such we submit the following select list of Hyacinths, Narcissi, and Tulips, from which the kinds wanted for either purpose may be ordered. In doing this it is advisable to state whether the bulbs are to be grown in earth or in water, as in the latter case such bulbs are chosen as are of most mature growth.

HYACINTHS.

Double Red.

Amelia, *red.*

Augustus Rex, *red.*

Aurelius Prudens, *red.*

Beauté Hortense, *red.*

Catharina Victorieuse, *rosy.*

Duchesse de Parma, *red.*

Euterpe, *red.*

Flos Sanguineus, *red.*

Gloria Solis, *red.*

Groot Vorst, *rosy.*

Hirsilia, *red.*

Il Pastor fido, *rosy.*

Julia, *rosy.*

La Beauté Supreme, *red.*

L'Honneur d'Amsterdam, *red.*

Madam Zoutman, *red.*

Marquis de la Coste, *red.*

Regina Rubrorum, *red.*

Rose Mignone, *rosy.*

Soleil Royal, *red.*

Superb Royal, *red.*

Waterloo, *red.*

Double Blue.

Alamode, *dark.*

Admiral de Ruyter, *dark.*

Aristides, *pale.*

Bleu Foncé, *dark.*

Bucentaurius, *dark.*

Comte de Veri, *dark.*

Comte de St. Priest, *pale.*

Duc d'Angouleme, *pale.*

Duc de Bronswick, *dark.*

Duc de Normandy, *dark.*

Florus, *dark.*

Georgius Tertius, *pale.*

La Majestueuse, *dark.*

Mignonne de Dryfhout, *pale.*

Monarque de France, *pale.*
Mon Amie, *dark.*
Montigni.
Orondates, *dark.*
Parminio, *dark.*
Pasquin, *pale.*
Passetout, *dark.*
Pourpre de Tyr, *dark.*
Susanna Elizabeth, *dark.*
Velours Noir, *dark.*

Double White.
Alamode, *white.*
Altesse Royale, *purple eye.*
Anna Maria, *purple eye.*
Blanchard, *white.*
Blanchfleur, *white.*
Comtesse de Dagenfeldt, *red eye.*
Duc de Valois, *white.*
Gloria Florum Suprema, *red eye.*
Grand Monarque, *red eye.*
Madam de St. Simon, *red eye.*
Mignonne de Delft, *red eye.*
Og Roi de Basan, *red eye.*
Spheri Mundi, *purple eye.*
Sultan Achmet, *white.*
Virgo, *white.*
William Frederick, *red eye.*
Duc de Berri d'Or, *yellow.*

Single Hyacinths.
Epichaus, *red.*
Henrietta Wilhelmina, *red.*
Konings Jewell, *red.*
L'Eclair, *red.*

Paix d'Amiens, *red.*
Emilius, *blue.*
Emicus, *blue.*
King's Mantle, *blue.*
La Crepuscule, *blue*
Lord Liverpool, *blue.*
Nimrod, *blue.*
William Tell, *blue.*
Madam Talleyrand, *white.*
States General, *white.*
Triumph Blandina, *white.*
Vainqueur, *white.*

NARCISSI.
Baselman Major.
Grand Monarque.
Grand Primo Citronier.
La Belle Leigeoise.
Primo Luteo.
Soleil d'Or.
Double Roman.
Single Paper White.

Narcissi are generally ordered by the half dozen or dozen.

TULIPS.
Clarimond.
Duc Van Thol.
Double Duc Van Thol.
Plus Aimable.
Pottebakka, *red.*
Pottebakka, *yellow.*

Select Tulips are generally ordered by the dozen.

We subjoin a list of handsome flowering Hyacinths which do not exceed 1s. 6d. per bulb: many
of these, though less new or rare, rank in regard
to absolute beauty, as highly as others which
cost twice, thrice, and four times that sum.

Perruque Quarrée, *rosy.*
Rouge Charmante, *red.*
Azure Incomparable, *pale blue.*
La bienne Aimée, *dark blue.*
Hugo Grotius, *rosy.*
Phœnix, *red.*
Constantia Elizabeth, *white,
    purple eye.*
Don Gratuit, *white, yellow eye.*
Duchesse de Bedford, *white.*
Gekroon Jewel van Haarlem,
    white, red eye.
Grande Magnificence, *white,
    yellow eye.*

Soleil brillant, *red.*
Beauté charmante, *blue.*
Diadem de Flora, *red.*
Violet Superbe, *white, purple eye.*
Chrysalora, *white, yellow eye.*
L'Or Vegetable, *white, gold eye.*
Ophir, *white and gold.*
Aimable Rosette, *red.*
Bishop of Munster, *blue.*
Boronicus, *blue.*
L'Ami du Cœur, *blue.*
Rouge Charmante, *red.*
Grandeur Triomphante, *white.*
Mon Bijou, *dark blue.*

We next enumerate the other bulbs usually
imported and forced, chiefly in pots of earth.

JONQUILS.
Largest Double Dutch.
Second Size ditto.
Large English.
Sweet-scented Single.
*Jonquils are generally ordered
by the dozen.*

IRIS.
Persian.
Chalcedonian.
Spanish Bulbous.
English ditto.
Peacock, large.
*Irises are generally ordered by
the dozen.*

ANEMONES.

Finest mixed double.
Fine   ditto   ditto.
Good   ditto   ditto.
Early single.
Double Hortensis.
Single   ditto.
Purple   ditto.

Anemones are generally ordered by the pound, but the three last sorts by the root.

RANUNCULUS.

Superfine mixed double.
Fine   ditto.
Good Border.
Semi Double.
Scarlet Turban.
Yellow Turban.
Ordered by the hundred.

CROCUSES.

Large Yellow.
Second Size Yellow.
Blue.
White.
Scotch.
Violet-striped.
Cloth of Gold.
Dutch ditto.
Saffron.
Dutch mixed.

English mixed.
Crocuses are ordered by the hundred.
Double Snowdrops, *ordered by the hundred.*
Fritillarias, mixed.
Ditto in sorts.
Colchicums, double.
Ditto   single.
Dog's Tooth Violets.
Stars of Bethlehem.
Gladiolus.
Guernsey Lilies.
Belladonna ditto.
Jacobea   ditto.
Martagon   ditto.
Tiger   ditto.
Pancratium maritimum.
All the above are ordered by the dozen.
Amaryllis longifolia.
———— Regina.
———— vittata.
Ixia crocata.
Ordered by the root.
Grape Hyacinths.
Large Musk ditto.
Small Musk ditto.
Double tuberose.
Tigridia pavonia.
Ordered by the dozen.

The above lists are taken from the annual Catalogue of Warner, Seaman, and Warner, Cornhill, who, with Mason of Fleet-street, are the principal importers of bulbs in the City of London.

If there is room in the villa green-house, a pot of sweetbriar, previously forwarded in a pit, hot-bed, or warm kitchen till the buds are ready to break, may be admitted for the sake of perfuming the air.

The same may be said of a pot or two of the moss or Provence rose, of the Persian lilac, and scarlet thorn ; but unless there is abundance of room, and ample means of forcing them, so as not to set them in the green-house till the bloom is expanded, they are much better kept out. So much winter brought into view tends to destroy the character of the place, and render it little better than a shrubbery.

If pinks, sweet peas, violets, and other showy and odoriferous flowers, can be forced in pits, so as to be in bloom before they are brought to the green-house, a few of them may be admitted ; but, as before observed, unless the green-house is very large, the fewer the better. A lobby, ante-room, stair, hàll, or even the drawing-room, are fitter places for forced productions than the green-house, where by contrast with the proper inhabitants they lessen the dignity of the place.

A few pots of mignionette, both of the dwarf and tree kind, should be kept in the green-house all the year round.

To have mignionette in perfection throughout the year, three sowings are requisite ; the first in the beginning of August for blowing in December, January, and February ; the second about the end of August for March, April, and May ; the third about the end of February for June, July, and August ; and the fourth about the end of May for September, October, and November. Sow in pots about 3 or 4 inches diameter, and when the plants have three leaves thin them out to four or at most five in a pot. Protect them from heavy rains by a frame, and from frost by coverings over the frame, and in the summer months keep the young plants under the shade of a wall.

The tree mignionette, which is merely a variety of the dwarf sort, requires exactly the same treatment, excepting that only one plant must be left in a pot, and this trained to a single stem. As mignionette when trained as a tree lasts much longer than when left to trail on the ground, one sowing in spring will produce plants which will last a year. In training, care should be taken to pinch off all blossoms that appear before the stem has attained the proper height. By pinching them off during a whole summer, the plants will be two or three feet high by the autumn and will make a fine appearance, and be profusely covered with bloom during the winter.

April, May and June.

Mignionette and hyacinths as during the preceding months; and perhaps a pot or two of Neapolitan violets (a variety of *Viola odorata* so named), of spring cyclamen, of auricula and choice wall-flowers or stocks. But the proper greenhouse plants will now make such a fine appearance that few of these extraneous articles will be necessary.

June, July, August and September.

During these months the green-house plants in pots are generally out of doors; and on the greenhouse stage are generally scattered a few pots of tender annuals, as balsams, cockscombs, globe amaranths, &c. In some cases a few pots of useful plants are very properly added; as of capsicums, egg plants, love apples, basil, and cucumbers trained to fan-shaped trellises fixed to the pots. All of these articles are admissible; the cucumbers, love apples and capsicums are useful in every family for pickling, stewing, and sauces, and they correspond well with the clusters of grapes which ought to be hanging from the rafters.

All these plants must be raised on heat; and the true way to bring them to a large size is, to use very rich soil, to plant at first in small pots, and every ten or twelve days to transplant them into pots a size larger, till at last the cucumbers are in pots a foot in diameter; the balsams into pots 8 inches in diameter; and the others into pots from 8 to 6 inches across, according to their natural size, &c.

October, November and December.

In the beginning of October the green-house plants are generally replaced ; but a few of the most hardy sorts, not above a foot or 18 inches in height above the pots, may be kept out two months longer in common frames, to make room on the green-house stage for a small collection of Chrysanthemums and Dahlias. These flowers, the Chrysanthemums especially ,produce a charming effect in the green-house at this season, when so few plants are in flower; and, with the exception of one or two sorts, it is only in the green-house that they will flower freely and maintain their blossoms in due form and colour. In the open air, at this season, the cold prevents the buds from expanding properly, and the wet drenches and disfigures the flower as soon as it is expanded. Whoever therefore would enjoy the Chrysanthemums must either force them in spring, in a pit or hot-bed, so as they may flower in the open air in August and September, or blow them in a green-house or in some other plant habitation.

For the Dahlias a green-house is not essential, as they flower earlier in the season; but by not planting a few of the roots till the end of May, and then putting them in pots, a few plants may be obtained, which will come into flower in October and November in a green-house, and make a fine assemblage along with the Chrysanthemums.

The varieties of the Chrysanthemum (*Chrysanthemum indicum* of Linnæus) have all been obtained from China within the present century, and chiefly within the last seven years. They are, the

Purple.	Quilled flamed yellow.
Changeable white.	Quilled pink flowered.
Quilled white.	Early crimson.
Superb white.	Large quilled oranges.
Tasselled white.	Expanded light purple.
Quilled yellow.	Quilled light purple.
Sulphur yellow.	Curled lilac.
Golden yellow.	Superb clustered yellow.
Large lilac.	Semi-double quilled pink.
Rose or pink.	Semi-double quilled white.
Buff or orange.	Small yellow single.
Spanish brown.	

All these varieties are beautiful; and, as the plants are so exceedingly easy of culture, we recommend one or two specimens of each of them to be grown in small pots. They are propagated by dividing at the root, or by cuttings taken either from the flowering stems or from the ground shoots. The cuttings from the stem make the neatest and smallest plants, and come soonest into flower: those of the root also make excellent plants, as do single root shoots or suckers slipt off with a heel and a few fibres attached.

To preserve a collection in little space, the best way is to grow the plants in single pots of a small size (say 2 or 3 inches diameter), and never to allow more than one stem to each pot. Take off the cuttings in May, and plant each in the centre of a pot 2 inches in diameter within, in rich loamy soil; place them in the shade, or in a cold frame covered with a mat, for a few days till they have struck, then remove them to a cool shaded airy situation, setting them on boards, slates, or bricks, or otherwise so placing them that the worms may not get into the pots. Here let them remain without shifting into larger pots or renewing the soil, but only supplying water (and manure water may be used if at hand), and tying them with black threads to neat rods about 2 feet or 2 feet 9 inches in length. By the middle of September the plants will be from 18 inches to 2 feet in length, according to their kinds, and clothed with foliage, and some of them with side shoots from the bottom upward. About the beginning of October the plants will show flower buds: in order to have strong flowers, pinch off all these but the centre one and three side ones; or, if the plant is not very strong and branchy, leave the centre one only: next, as the plants have now attained nearly their full height, adjust the props so as they may just reach within an inch of the top bud, and no higher or lower unless the plant is very vigorous, when it may be a few inches lower. These props ought to be made tapering, and should terminate in points not more than 1-16th of an inch in diameter. Tie the stem and side flowers neatly and symmetrically to them with small black threads; clean the pots, stir up and freshen the mould on their surface, and remove them to the green-house. Here for the first week let them have all the air possible night and day; the second week the air may be diminished, and afterwards they will bear the usual treatment of green-house plants at this season.

There are various other modes of growing the Chrysanthemum, but the above is by far the best with a view to flowering them in the green-house.

The Dahlia is a plant as easily cultivated as the common potatoe, and, like it, the chief attention it requires is to keep the roots from being injured by frost. For this purpose they are taken up yearly in autumn, and kept in dry sand in a cellar or shed during winter. In spring, those intended to flower early are planted in pots in

February, and forwarded in a pit or hot-bed, and when grown one foot or more, by the end of May they are planted in the open air, either turned out of the pot or the pot buried in the soil. Those meant to flower later are planted in the potatoe season in May in the open ground; and those intended to flower late in the green-house are reserved in dry sand till the end of May or beginning of June, and then potted and plunged in the open ground. When they come up, limit the stems to one or three, and tie them neatly to rods of 3 or 4 feet in length. They will throw out side-shoots, which require to be thinned when they are very numerous; and at no time should more than one flower-bud be allowed to come forward on each side-shoot. About the beginning of October their future beauty and effect may be predicted, and now they ought to be nicely adjusted to the props, a bud left on each side-shoot, and three or four on the top of the main shoot. Remove them to the green-house, and treat them exactly as the Chrysanthemums.

The Dahlia requires a pot at least 6 inches in diameter even to flower one shoot freely, and therefore only a very few can be admitted into a small green-house. These few may be the following, or some of the endless sub-varieties which belong to them.

K

Of the fertile-rayed (*Dahlia superflua*), the purple, rose, pale, white, sulphur, yellow, tawny, copper, brick-red, dark red, pomegranate-coloured, dark purple. Add a few of the double and semi-double sorts for variety, though single flowers are preferred.

Of the barren-rayed (*Dahlia frustranea*), the scarlet, orange, and saffron: of this species the double varieties are the most beautiful; whereas of the fertile-rayed species, the single flowers are considered more beautiful than the double.

Sect. V. *Selection of Plants proper for a Conservatory, or Green-house, in which the plants are not grown in pots, but in beds and borders.*

The next point to consider is the stock proper for a conservatory or green-house, in which the plants are not grown in pots, but in beds and borders.

The first observation we have to make here is, that in such a green-house, unless of very great extent, the collection must be very limited, because the species cannot be kept in such small compass as when in pots; nor will they flower unless allowed to acquire a considerable size. Hence it is, that the stock proper for a conservatory consists chiefly of the larger-growing genera, as Camellia, a few of the larger Heaths, Protea, Banksia, Acacia, Myrtus, Melaleuca, Malva, Crotalaria, Metrosideros, Colutea, Cassia, Buddlea, Nerium, Struthiola, Lantana, and especially Citrus. In short, oranges, lemons, camellias, myrtles, banksias, proteas, acacias, melaleucas, and a few other Cape and Botany Bay plants, are all that can with propriety be admitted in a small conservatory:—there they must have abundance of room, air and light; and, as we have before observed, the free enjoyment of the weather during the summer months.

A conservatory, therefore, can never become so general or so desirable an addition to a villa as a green-house. It is a splendid accompaniment to a mansion or palace, and for such edifices alone is it suitable.

The following is a detailed selection of Conservatory Plants.

[L. *largest size*; S. *smallest size*. *Those having no mark are intermediate between* L. *and* S.]

Acacia armata, L.	Acacia alata
A. undulata	Aotus villosa, S.
A. verticillata, L.	Araucaria excelsa, L.
A. longifolia, L.	A. imbricata, L.
A. nigricans	Azalea indica, *single red*, S.
A. pulchella	A. indica, *double purple*, S.
A. pubescens, L.	A. indica, *single wh'te*, S.
A. decurrens, L.	Banksia ericifolia, L

Banksia, *all the genus*

Beaufortia decussata

Boronia pinnata

B. serrulata

Brachysema latifolium ⎞
B. undulatum ⎠ *espalier*

Callistachys lanceolata, L.

C. ovata, L.

Calothamnus quadrifida, S.

C. villosa, S.

C. gracilis, S.

Camellia, *any species or variety.*

C. Sasanqua *will do well to run up a pillar or train upon an espalier.*

Chrysanthemum indicum *might be introduced when in bloom, and plunged in the borders as if growing there.*

Citrus, *any or all according to the size of the house and the taste of the owner.*

Clethra arborea, L.

C. variegata, L.

Corræa speciosa, S.

C. viridis, S.

Crotalaria elegans, S.

Crowea saligna

Cytisus canariensis

C. proliferus

C. foliolosus

Dais cotinifolia, L.

Daphne odora

Datura arborea, L.

Daviesia ulicina

D. latifolia

Dryandra, *all the genus, particularly* floribunda *and* longifolia.

Epacris grandiflora

E. pungens, S.

Eutaxia myrtifolia, S.

Fuchsia coccinea (*espalier*)

Gastrolobium bilobum, S.

Gnidia pinifolia, S.

Goodia lotifolia

G. pubescens

Grevillea linearis

G. buxifolia

Hovea Celsi, S.

H. linearis, S.

Humea elegans *might be turned into the conservatory when coming into flower; grows nearly six feet in height.*

Indigofera australis

Lambertia formosa, S.

L. echinata, S.

Loddigesia oxalidifolia, S.

Magnolia fuscata, L.

Melaleuca splendens

M. hypericifolia

M. diosmifolia

M. thymæfolia

Metrosideros floribunda

M. rigida

Nerium, *to be brought into flower in the stove, and then plunged in the borders of the conservatory.*

133

Pelargonium, *any of the shrubby species for a year or two only.*
Pimelea linifolia, S.
Pinus lanceolata
Pittosporum undulatum, L.
P. Tobira, L.
Polygala speciosa
P. cordata
P. oppositifolia
P. myrtifolia
P. latifolia
Polygala bracteolata, S.
Prostanthera Lasianthus
Pultenæa daphnoides, S.
P. stipulacea, S.
P. retusa, S.
Rosa Noisettea
R. odorata
Stenanthera pinifolia, S.
Tristania laurina, L.
Zieria Smithii, S.

Climbers, Twiners, &c.

Acacia prostrata
Bignonia grandiflora
B. capreolata
Billardiera scandens
B. mutabilis
B. longiflora
Cobæa scandens
Clematis florida
Convolvulus canariensis
Coronilla juncea
Cryptostegia grandiflora
Dolichos lignosus
Glycine Comptoniana
G. maculata
G. macrophylla
G. sinensis
G. rubicunda
Hibbertia dentata
H. grossularifolia
H. volubilis
Hoya carnosa (*in a warm corner*)
Lonicera japonica
L. sinensis
L. flava
L. coccinea
Maurandia semperflorens
M. antirrhiniflora
Passiflora cærulea
P. fimbriata
P. cærulea racemosa
Pelargonium scutatum
P. ————————carneum
P. ————————superbum
P. pinguifolium
P. zonale, *many varieties*
P. peltatum
Rosa Boursaulti
R. multiflora
Rubus rosæfolius

In addition to the above, which are plants of known beauty, we would recommend for the conservatory, as for the green-house, some of the newly introduced plants which have not yet flowered or been named. Such plants are calculated to excite great interest, from the circumstance of their being entire strangers; and when any such plant happens to flower for the first time in Britain, in *our garden*, there is some satisfaction in seeing the circumstance noticed in the botanical works where it may be figured, described, and named.

Many plants well suited for this purpose may be obtained in Mr. Mackie's Nurseries in the King's-road, and at Clapton; the collectors whom that enlightened and spirited cultivator maintains in Australasia, and other parts, having been singularly successful. Mr. Tait of the Sloane-street Nursery has many of the new plants brought from Mexico by Mr. Bullock: Messrs. Brooks have imported a number of novelties: Messrs. Loddiges receive the contributions of Protestant Missionaries from every quarter of the world: and in short, in the present day, such is the taste for botany and gardening, that almost every Nurseryman has something which nobody has but himself.

SECT. VI. *Of the placing or arrangement of the Plants in a Green-house.*

The stock for the green-house being got together, the next thing is to dispose of them on the stage in the best manner. One common rule is, to place the smaller plants in front, and the larger behind; this is most generally proper, because it brings the smallest plants nearest the light. In some cases however, where there is a front wall which reaches higher than the lowest shelf of the stage, or which throws a shadow on it, it is better to place the smallest plants three or four shelves high on the stage, and plants of a larger description on the bottom shelves. This more effectually attains the object of getting the smallest plants nearest the light, and also brings them near the eye.

A second general rule is, to mix all the different plants as thoroughly as possible, so that no two of a species, or even if possible of a genus, may be seen together. The object of this mixture is to produce variety; but a little reflection will convince any one, that instead of variety it produces a sameness of mixture exactly the reverse. Variety requires a certain degree of distinctness of character or feature, on which the eye can repose itself before, and then proceeding to another : but where every thing is indiscriminately mixed together, there can be no features, nothing on which the eye can dwell with satisfaction ; all is confounded, and reduced to a mere chaos of forms and colours. Instead of this mode of mixture, we recommend, as much as is practicable, that each genus and species be kept by itself; and where a number of genera form a striking natural order, as the *Ericeæ*, *Geraniaceæ*, &c., that the whole be grouped together ; not in a formal manner, but so as to show a sort of relationship or connexion, and at the same time to blend in with surrounding genera. It may be alleged, that where some of the species of a genus are large plants, and others small, this could not well be done consistently with showing off the plants to advantage : but such an objection arises from taking too formal a view of the subject. By keeping plants together, it is not meant to keep them in contact in a compact clump, but to place them in visible connexion in irregular groups, which is quite consistent with placing the tallest plants of the group on the upper part of the stage, and the lesser plants nearer the spectator. It is sufficient that the connexion of the species be recognised by the eye, and that there appear in the green-house, what there always is in natural scenery, something like a natural gradation, and blending of character, in shrubs, trees, and plants growing together.

Besides some attention to keeping together genera and species, or natural orders where the case admits of it, it is desirable also to see that they harmonize in growth with the adjoining genera or natural orders. For example, after the group of Camellias, it would be too violent a contrast to commence with Succulents; but the Myrtle tribe, or *Proteaceæ*, might very well adjoin; then *Leguminoseæ, Geraniaceæ*, and next the Succulents. In short, the most perfect mode of arranging the plants on a green-house stage would be to follow the natural system of Jussieu : but as to do that completely is impracticable, the next thing is to follow it as far as we can — and the third best thing, to keep the genera together. Neglect of the last principle as a general guide is quite inexcusable even in the smallest green-house,—attention to it will be amply repaid by the effect produced even on a chamber flower-stage : as a proof of this, we have only to refer to the green-houses of nurserymen who are extensive growers of green-house plants, and especially to the well-arranged collections of Messrs. Loddiges in their garden at Hackney.

Let it not be thought, however, because we recommend every genus to be kept by itself,

that we carry this so far, or hold the principle in
so arbitrary and absolute a manner, as to exclude
all or any good effects that may result from a deviation from it. No man of sound sense ever holds
any principle in so absolute a manner. That
good effects may and do arise from a deviation
from general principles, we readily allow, and on
all fitting occasions would gladly avail ourselves
of them. For example, in the midst of a broad
group of heaths, it may produce a striking effect
to place a tall handsome camellia, or to distribute
two or three hyacinths and polyanthuses in flower :
a tall plant of tree mignionette may have a fine
effect among the succulents, and so on. The
effect produced always carries its own argument,
and justifies the deviation from the general principle ; for extraordinary effects, in fact, only become
such by being beyond the reach of ordinary rules.

The third common rule in placing plants on a
stage is, to adjust their heights so accurately as
to make them dress off from the floor to the top of
the back wall in one even surface of verdure, like
a shorn hedge. This produces a striking effect
at the first glance, but is unfavourable to a prolonged interest, by a more detailed examination,
of whatever species the collection may be composed ; only their tops are to be seen, and no
flowers but what issue from the points of the

shoots. What we would recommend is, to make the sloping surface much more irregular; by which means the eye of the spectator would examine the sides of many of the plants as well as their tops, and the plants themselves would receive more benefit from the light and air. Any mode in which plants can be placed on a stage has a tendency to draw them up to slender forms, naked below; but when they are crowded, and so nicely adjusted to an even slope as is commonly aimed at, the deformity is greatly aggravated. This evil is beginning to be felt by the most eminent exotic gardeners, who now place their plants much thinner than formerly, and so adjusted in regard to size, that the direct rays of the sun, in many cases, repose on the earth in the pots. The green-house of the Comtesse de Vandes at Bayswater, affords a fine example of the beneficial effects of this sparse mode of placing the pots. The plants there, are generally so well clothed with foliage from the pot upwards, that almost any one of them taken at random, and placed on a pedestal, would form a fine single object. This, indeed, is the true test of success in the culture of plants in pots.

With respect to the arrangement of plants in a conservatory, the general rules are the same as for the green-house; viz. to place the tallest-growing sorts behind, or towards the back wall or north side of the house, so as to present a sloping surface to the sun. This surface is also best adapted for meeting the eye of the spectator in the front path, while in the back path he is under the shade of the higher trees and shrubs.

But it very frequently happens that a conservatory is placed south and north with a span roof, a walk round the whole, and one up the middle. In this case the higher-growing plants are placed along each side of the middle walk, and lesser ones sloping towards the side walks. The creepers are planted on the props which support the span roof, and which form a sort of groined arcade over the middle walk : this is the situation in which creepers do least injury; and in short this plan, all circumstances considered, is the best for a large conservatory.

It is a desirable circumstance in a conservatory, to mix evergreens and deciduous sorts together (though of the latter, only the most beautiful flowerers should be introduced); and when this is done, the evergreen shrub should always be contrived to cover the surface under the deciduous tree : at any rate, the contrary practice should never be adopted ; for nothing can be more opposite to natural principles, than for a deciduous tree to grow in the shade or under the drip of an evergreen. By trees, of course, we are to be considered as referring to such woody plants as come up with single erect stems, as the orange, in opposition to such as come up with bushy stems, or numerous suckers, as the myrtle, rose, fuchsia, &c.

Wherever there is a tolerable collection of green-house plants, it is desirable that they should be named. To make use of the proper and correct names of objects, is an important part of common conversation, and nothing can lead to a correct application of botanical names, but affixing them by some means to the plant, or pot containing it. It is also very desirable that plants should be named, with a view to children, as the perusal and recollection of these names will aid in strengthening their memory, enlarge their powers of pronunciation, and create a taste for natural history.

The most common mode of naming plants in pots, is to take a flat slip of wood, sharpen one end ; rub a little white lead on an inch or two of its smoothest surface at the opposite end, write the name with a black lead pencil on this lead, and then insert the stick in the pot. This is the mode adopted in the nurseries, and will last several years. Another suitable mode for a private green-house, is to put the name in black letters of oil colour on a neat small stick painted white : but the most recent and pleasing mode is that of using earthenware as a substitute for wood, and either writing the name with common ink, pencilling it with black or common lead, or painting it in oil. Such naming instruments are now made in abundance at the Potteries, and are to be had at Spode's, Wedgwood's, or any of the principal earthenware shops in the Metropolis and large towns. They are in use at Messrs. Loddiges, Sion-house, and in the hot-houses of the Horticultural Society.

A small book containing a list of the names should be prepared, and occasionally the stock examined by it, to ascertain what sorts may have been lost, &c. A very agreeable and instructive task for a young person studying plants, would be to turn this list into a descriptive history, with a graphic sketch of every plant appended.

CHAPTER III.

OF THE GENERAL CULTURE OF GREEN-HOUSE PLANTS, AND THE MANAGEMENT OF THE GREEN-HOUSE.

THIS subject embraces a variety of particulars and operations, which have for their end the growth and flowering of the plants, their maintenance in health, and the general beauty and effect of the green-house. We shall first premise a few remarks on the stock of materials requisite for culture, and next consider in succession the general principles of exotic culture, and the culture and management suitable for the four quarters of the year.

SECT. I. *Of forming a stock of Soils, Pots, Props, and other articles, for Green-house culture.*

It may be useful to commence this section by stating a general principle as to soils, which it would greatly benefit the practical gardener to keep constantly in his mind. All plants that are susceptible of much alteration by culture, will grow in nearly the same soil. All plants that by culture have their parts enlarged, and rendered more succulent by pulverizing the earth to a considerable depth, draining it if over wet, removing other plants whether weeds or of the same species, sheltering by choice of low situation or otherwise, and manuring and pruning;—all such plants, we repeat, as are much affected by these and other operations of culture, will grow in a sandy loam. This, it will be at once seen, includes nine-tenths of the perfect plants (we exclude *Cryptogameæ*); and of this nine-tenths, it may be safely affirmed that they will all grow well in a sandy loam. They will not all attain equal perfection in this soil; for some may require it to be a little stronger, and others a little lighter than the general average; but they will all grow better than if left to themselves, in their natural soils and situations. Of the remaining tenth of the perfect plants, it may be said that they are absolute as to soils, the greater part being either aquatics, or plants that grow in what are called aquatic soils, that is, peat earth plants. There are only a few plants which grow on sands and rocks so absolutely as not to be made to grow in sand with a little loam.

If this theory be correct, it will follow that the chief earths required by the cultivator are loam and sand, from which he will be enabled to form composts for all plants of culture; and next peat earth, which, with the aid of sand, will enable him to form composts for all plants of unalterable habits as to soil.

What loam is, and how, where, and when it is to be procured, are the next subjects which we might treat of; but this we fear we could not do to much advantage. Every gardener knows a loamy soil from a sandy, gravelly, limy, peaty, stony, or clayey soil at first sight; and it would not be easy to communicate this fact verbally to a person who only knew soils by reading about them. It may be sufficient to say, that all nursery gardeners choose it for the scene of their operations when they can, and that it is finely exemplified in most of the London nurseries west of the Metropolis. It can no where be better studied than in the vineyard at Hammersmith, or in the garden of the Horticultural Society.

The gardener knows how to collect loam for green-house plants, by paring off a few inches of the surface of a loamy grass field; or taking in a body the earthy materials of an old hedgerow standing on a loamy soil. This should be carted to a convenient part of the garden, laid in heaps or ridges, and turned over two or three times a year, so that every part may be heated by the sun, moistened by the rains or snows, and frozen by the winter's frosts.

Peat soil is to be collected chiefly from peat

bogs; but also from thin dry peaty surfaces, which are found occupying the low parts of wastes and commons, and sometimes woods. Wherever aquatic vegetable matter has accumulated and ceased to grow, there peat may be dug, either in surface turves, or in spitfuls from under the surface; and these collected and laid in heaps will gradually decay, and moulder into peat earth. At present there are few districts of country so entirely subjected to aration, as not to admit of situations where peat earth may be found; but the time may come, when this material may have to be imported from the north of Scotland, or Ireland, or even from other states in the north of Europe. But before such a scarcity occurs, it will be found that leaf-mould, especially where the leaves of the pine and fir tribe are used, forms an excellent substitute in most cases.

Sand for the purpose of mixing with loam or peat, or for being used alone in striking the more difficult cuttings, may either be obtained from pits, rivers, or by pounding sandstone. The siftings of gravel are generally unsuitable, as earthy and containing a good deal of oxide of iron, known by the brown or yellow colour of such sand. Good sand is generally white, or but slightly tinged.

Besides loam, peat, and sand, some rubbish of old buildings will be occasionally wanted for succulents, though sand answers nearly if not entirely as well. This material is best obtained when wanted; for when kept in a compost ground for a year or two, and turned over with the other soils, it assumes the character of a limy earth.

Leaf-mould or vegetable mould may be considered more in the light of a manure than of an earth. It is formed by collecting leaves together in heaps, carrying them to the compost ground, or space in some inconspicuous part of the garden devoted to moulds, dungs, and other stock articles, and there laying them in ridges, and turning them over three or four times a year. It is natural to suppose that the nature or kind of the leaves collected will have a material influence on the mould produced; it is not however ascertained to what extent this is the case. Speechly collected oak leaves chiefly as requiring longer time to decay, and for that reason being more suitable to use as a substitute for tanner's bark, as a fermenting vegetable substance to produce bottom heat: M'Phail collected indiscriminately whatever leaves came in his way, and he states that the sorts included all those found in modern shrubbery plantations; they fermented well, and produced mould in which he grew the cucumber, and, with some addition of loam, the melon and pine, to the greatest perfection. M'Phail is the only writer

who acknowledges that there were leaves of the
pine and fir tribe among those he collected; but
the influence of these, he says, if different from
that of the common broad leaves of deciduous
and evergreen trees, he could not discover.

From observing the vegetable mould in old
pine and fir forests, we are of opinion that this
mould is of a different quality from that of the
leaves of oaks and other non-resinous plants; and
we think it might well merit the gardener's attention to select a quantity of this mould, keep it by
itself, and try whether in the culture of heaths and
terebinthinate plants it would not be a good substitute for peat.

Mould of rotten dung is another requisite in
the culture of green-house plants. This is easily
formed, by bringing together, in a heap or ridge,
a quantity of old hot-bed dung, or spit dung of
any sort, and turning it over occasionally for a
year or more till it becomes a soft black mould.
There are three varieties of mould of dung which
ought to be obtained if convenient, and in the
compost ground ought to be kept quite distinct.

The first is the mould of horse- or stable-dung,
or that procured from old hot-beds; this is reckoned the strongest, and when added to earths is
commonly looked on as a manure or enricher
more than as an earthy ingredient.

The second is the mould of cow-dung, which as a manure is very weak, and is deemed excellent for entering into composts preparing for bulbous roots, and even in many cases as substitutes for peat earth.

The third is the mould of rotten straw, such as thatch, packing refuse, cleanings up of rick-yards, coverings of beds, borders, or mushroom ridges, or in short any straw that has not been mixed in any way with animal matters, as urine, excrement, &c. This forms the best substitute for peat earth; and for the great majority of purposes for which that earth is used, this mixed with a proper proportion of sand will be found eligible; mixed with finely sifted rotten tan, sand, and a little loam, it forms a soil for hardy American plants, in which they thrive as well as in peat or bog earth.

A sort of vegetable mould, which we have seen collected and used with the greatest success, is the following: Form a large heap of spray of trees, as clippings and cutting of hedges, old pea-sticks, prunings of trees and shrubs, and any other small shoots or branches chopped into lengths not exceeding a foot or eighteen inches; add whatever of saw-dust and rotten tan can be got, and if circumstances are favourable, pea, bean, and other haulm, stems of cabbages and brocolis, and other garden refuse; add further turves from any grassy surface not a clay—if peaty turves so much the better, or of the tussocks of meadows, and such tufts of grass and grassy roots as are found by ditches: add any thing else of the vegetable kind than can be got: then, when all is collected that it is thought can be got together at the time, turn the heap over, and mix the whole well together. This done, get an equal bulk of fresh stable-dung and mix along with the heap, in the form of a bed, either to be covered with earth and a frame, or with earth and hand-glasses. An excellent hot-bed will thus be produced the first season; and in the second, if the top earth, in which the plants were grown, and dung, and composition are all mixed together and turned over once a month or six weeks, then the third year there will be produced an excellent vegetable mould fit for any purpose, but particularly suited for American plants.

Quick-lime is sometimes wanted in green-house gardening, for forming lime-water to destroy worms and other vermin, and to wash the flues and walls in some places; but as this article cannot long be kept fresh, it is better to get it as wanted from the dealers, than to lay in a stock of it.

Scoria or smiths ashes, soapers' waste, or any vitreous gravelly matter, will be wanted for form-

ing a platform impervious to earth-worms, on
which to set the plants during the time they are
out of doors in the summer months. If wanted,
however, it is chiefly on the first formation of the
platform, and a small stock for annual repairs;
but it is found by experience that the most effectual mode, and the cheapest in the end, is to pave
the platform with flag-stones, or form a flooring of
broken bricks and Roman cement.

The next stock article which we shall mention
is pots and saucers. There should be a stock of
pots of all the sizes in common use, but chiefly from
2 to 6 inches in diameter as those most generally
adopted in the green-house; there should also be a
few saucers or flats for each of the different sizes of
pots, in case of taking any plant while in flower
into the living-rooms. Both pots and saucers must
be kept quite dry in a shed formed on purpose for
holding different stock articles, and for the operations of potting, propagating, cleaning plants, &c.

Fresh green moss for putting on the surface of
pots taken into the drawing-room; mats for protection, and for supplying ribbons for ties; fine
black flaxen threads for ties; fine wire for the same
purpose; neat tapered rods of all sizes painted
green to be used as props; small wire props,
nails, lists, a brush and sponge for cleaning
leaves, and a variety of other articles in common

use in gardening, should be at hand in moderate quantities.

Tobacco to be consumed in smoke for destroying insects: sulphur for strewing over plants attacked by the mildew : soft soap for washing off the scale : brushes for applying the last, and a fumigating bellows for the first, are also requisites.

There are no tools or implements peculiar to the green-house, but a syringe of the much improved form invented by Read, and known as Read's syringe; a thermometer, and especially Six's registering one, which shows the extremes of temperature that have happened during the night; bell-glasses for striking cuttings; naming-sticks and a little white lead; sheets of strong paper or pasteboard for temporary shade; and a variety of other articles common to almost every garden, will occasionally be brought into action.

Our enumeration of these articles may appear to some sufficiently formidable, and at first sight would seem to deter from the idea of keeping a green-house in a small garden; but in practice these articles occasion no difficulty or expense, those of consequence being peat earth and pots.

SECT. II. *Some general Maxims of exotic Culture.*

A plant in a pot, though conveniently circumstanced as to deportation, yet is in a highly artificial state, requires a highly artificial culture, and
is more liable to be injured or destroyed than one
growing in the common soil. A plant in a pot
which is placed on a shelf or stage and surrounded
by air, is much more liable to accident than one
plunged in the soil or surrounded by earthy matter, or even straw or leaves. The want of a steady
temperature and degree of moisture at the roots of
plants, is more immediately and powerfully injurious to them than atmospherical changes; earth,
especially if rendered porous and sponge-like by
culture, receives and gives out air and heat slowly;
and while the temperature of the air of a country
may vary twenty or thirty degrees in the course
of twenty-four hours, the soil at the depth of two
inches will hardly be found to have varied one
degree. With respect to moisture, every cultivator knows, that in a properly constituted and
regularly pulverized soil, whatever quantity of
rain may fall on the surface, the soil is never saturated with water, nor in times of great drought
burnt up with heat; the porous texture of the
soil and subsoil being at once favourable for the

escape of superfluous water, and adverse to its evaporation, by never becoming so much heated on the surface or conducting the heat so far downwards as a close compact soil. Now these properties of the soil relatively to plants can never be attained by growing them in pots, and least of all when these pots are so placed as to be surrounded by air. In this state, whatever may be the care of the gardener, a continual succession of changes of temperature will take place on the outside of the pot; and the compact material of which it is composed being a much more rapid conductor of heat than porous earth, those changes of temperature will soon be communicated to the web of roots which line its interior surface. With respect to water, a plant in a pot surrounded by air is equally liable to injury. If the soil be properly constituted, and the pot sufficiently drained, the water passes through the mass as soon as poured on it, and the soil at that moment may be said to be left in a state favourable for vegetation; but as the evaporation from the surface and sides of the pot, and the transpiration of the plant go on, it becomes gradually less and less so, and if not soon re-supplied, the earth would become dry and the plant shrivelled and liable to die, either from the want of water, or its sudden application. Thus the roots of a plant in a pot surrounded by air, such as the pots on a green-house stage, are liable to be alternately chilled and scorched by cold or heat, and deluged or dried up by superabundance or deficiency of water, and nothing but the unceasing care and attention of the gardener to lessen the tendencies to these extremes, would at all preserve the plant from destruction. Hence the advantage of plunging pots in sand, ashes, earth, saw-dust, tan, or any porous non-conductor; and also of shading them by leaves, straw, or moss; or, where this cannot be done, placing them on cool stone platforms, which do not admit air from below like open gradations of shelves.

A careful imitation of nature is not always the proper mode of treating plants in a state of art. It is obviously erroneous when applied to plants of culture, as to most culinary plants and fruit trees; but correct when applied to plants of fixed habits, as heaths, or plants whose characters and habits in a wild state it is desirable to preserve; as for example medical plants, which are all rendered less efficacious in proportion as they are affected by culture.

Though an imitation of nature may not be always the best mode of culture, yet no culture can be successful that is not conducted on natural principles. No culture can be successful which does not proceed on the value of roots and leaves, as the foundation of all the other productions of the plant, and reciprocally of each other's growth —on the value of light to the maturation of leaves —and of heat and moisture to the promotion of growth, &c.

The living principle in most plants ceases to exist when they are detached from the soil, and surrounded by air. The same may be said of the parts of plants artificially detached, as branches, shoots, leaves, &c.; but not of bulbs and tubers, which are entire plants in embryo, and, like seeds (which are the same thing), may be kept out of the soil for some time. The living principle in plants detached from the soil, and in shoots or detached leaves or flowers, may be preserved for a considerable period by lessening evaporation from their surfaces; as by inclosing in a box or case, packing in loose straw, &c.; and hence the value of a botanic box. Life may be maintained a still longer time, by inserting the roots or root ends of shoots or leaves in soil, or moistened moss, straw, or other matter that will supply a moderate degree of moisture. Thus it appears that plants when in a dormant state are not entirely so; but that there is going on a certain degree of circulation of sap, or waste of life, and which of course requires a supply.

All extraordinary stimuli applied to plants, as manures, heat at an unusual season, &c. in proportion to the effects they produce, leave a corresponding weakness when withdrawn. An extraordinary produce of blossoms, or fruits, in any one season, is commonly followed by a less than ordinary display in the season following.

The two greatest stimulants to growth are water and heat; the one should never be applied to any great extent without the other, and both should be used rather to second the efforts of nature, than to force her into activity. Thus, when plants are growing vigorously they should be watered freely; when slowly, sparingly; when dormant, very seldom; and in the case of bulbs and tubers in a dormant state, scarcely at all till shortly before their season of re-vegetation.

As water and heat are the greatest stimulants to growth, so light is the greatest stimulant to perfect that growth, and render it mature. Plants will grow for a time with water, heat, and air only, but not long; and their productions will never be of a green colour; nor will their blossoms or fruits ever arrive at the slightest degree of perfection.

The art of man can supply to plants every thing but light; therefore, in placing plants in artificial circumstances, light is the first object requiring attention. Plants, to enjoy the full benefit of light, must enjoy it directly from the sun ; light which has its rays much deranged by refraction or decomposition, as by a prism for example, will not produce the same greenth of leaves, brilliant-coloured blossoms and fruits, as light which has not passed through glass. But as all plants in green-houses can only enjoy their light, or at least a great portion of it, by receiving it through glass, hence the necessity of choosing the very best glass—that which is clearest and has fewest inequalities of surface, in order that the light may pass through it as little changed as possible. Ex-perience proves that the rays of light after passing through glass are considerably more decomposed, or at all events much more unfavourable to vege-tation, the distance of a yard or two from the glass, than close to its surface. Hence the great im-portance of placing plants near the glass.

Light being required for bringing the leaves of plants to perfection, it follows that plants which grow all the year, as some *Geraniaceæ*, must re-quire abundance of light all the year; but that others which grow only once a year, as camellias and oranges, may pass a part of the year with less light than when they are in a growing state. Nature acts in conformity to this law, and generally produces the growth of plants at a season of the

year when most light is afforded for maturation.
There are a few plants which grow perpetually;
but these are chiefly natives of the tropics, where
the light is nearly equal through the year. The
practical conclusion to be drawn from these
remarks is, that such green-house plants as are in
a growing state should, as far as is consistent with
other arrangements, be placed nearest the light;
and that those which have done growing, and
have matured their leaves, may be placed in situations much less illuminated than the others.

The habits of plants as to dropping their leaves
are different. Some part with the whole annually,
and at once, and leave the plant bare, as deciduous plants; others part only with a portion
annually, and retain a sufficiency to render them
perpetually clothed, as evergreens. The first
suffer no injury from being kept whilst in a naked
state in a situation with little or no light. Hence
some green-house plants, as the *Fuchsia*, *Aloysia*,
&c., may be removed from the stage in their rest-
ing-season, and placed under it, or in any dark
part of the house, or in a temperate shed or cellar.

Air (independently of its motion as wind) is
essential to every function of plants, from germi-
nation to ripening the seeds. A free circulation
of air is essential to the flavouring of fruits, and
in green-houses to the carrying off of damp un-

wholesome vapours, or excess of moisture in the atmosphere, which when not accompanied by abundance of light and heat, is apt to create a mouldiness on plants and the surface of their pots, and encourage the growth of mosses and fungi, which are doubly injurious to plants, by impeding transpiration and imbibing their nourishment. As from the moist state of the earth in the pots, and the warm temperature of the greenhouse compared with that of the open air during winter, its atmosphere must then be powerfully charged with moisture, hence the daily attention required to admit fresh air and promote its circulation.

As the presence of air is essential to the growth of plants, and of fresh air to their healthy state, so the action of wind is essential to their strength. Strength of stem is not necessary to all plants, for example, to creepers and climbers, which, whatever advantage they derive from air, benefit little or nothing from wind ; nor do they require to possess the strength of stem which wind gives, because nature has furnished them with the means of attaching themselves to other plants with strong stems. Wind produces in green-house plants what is called bushiness, or a close compact branchy form, which is always the best for a shrub in a pot, as a single stem with a head is the best for a small tree, as in the case of the orange. A salutary degree of wind is admitted to the green-house by opening the doors, and as much of the glazed covering and sides as possible, when it is gently windy in the open air; and the full effect of wind is enjoyed when the plants are set out in the open garden during the summer months. One suggestion of importance we offer as connected with the latter practice : it is, that when benefit is expected from the wind, the tall plants should not be tied to cross rods or lines by their tops, as is commonly the practice, but near the lower ends of their stems. By this means the wind is enabled to bring the upper part of the stem, that is all that is above the tie, into motion,—a thing essential to their benefiting by wind. Where the stem of a plant is fixed by the common mode of tying, the wind may invigorate the branches to such a degree as to render them too heavy for the slender stem from which they issue, and which, owing to the mode of tying, the wind cannot invigorate.

The habits of plants may be altered to a surprising extent. The most vigorous timber tree trained in the horizontal manner against a wall, like a garden pear tree, will not produce its massy trunk, as when exposed on all sides to the influence of the weather. A shoot of the common

M

pink, carnation, or sweetwilliam, if left singly on a root and trained upright to a prop, will form a not inelegant little evergreen tree, and the same may be said of mignionette, and some of the annual violets. Annual plants, if their flowers be constantly pinched off when they begin to appear, will become perennial, and perennial plants forced, often become annual or biennial.

Plants which have a natural tendency to propagate themselves by suckers, bulbs, runners, &c., are generally sparing in their production of seeds. Annual plants can seldom be readily propagated by other means than by seeds, and hence they are almost always fertile in their production. In all woody plants which propagate themselves abundantly by extension, it will increase the tendency to produce blossoms, to remove all suckers, and all shoots and branches near the ground, by which the whole of the vigour of the plant is thrown into its upper parts; and the upper parts of almost all plants are those where blossoms are first produced.

Where woody plants so treated are yet tardy of producing flowers, this tendency may be promoted by cutting off a ring of bark from the stem; or from any one of the branches of the plant, if the effect is meant to be limited to a branch. The season for performing this operation, is when the plant is in a growing state in spring : the ring or circle of bark should not exceed a fourth of an inch in width, in order that it may heal up at the end of the same season, or in the course of the next year. In green-house plants, the best part of the stem on which to perform the operation, is immediately under the surface of the soil, by which means the notch is unseen. The ring should penetrate to the wood, but no deeper.

The rationale of the operation of ringing is as follows :—The sap of plants, when they begin to grow in spring, rises from the roots to the upper extremities of the shoots through the vessels of the wood, and chiefly of the young wood : it is elaborated or prepared by the leaves, as the blood of man is by his lungs and liver, and then it is returned by the bark, nourishing each part as it goes along, and carrying down a large supply of nourishment to deposit in the root for next season. Now the ring intercepts the supply of nourishment on its way to the root, and obliges it to be deposited in the buds of the branches ; thus giving a great accession of force to them; and causing some or many of them to become blossom buds, which would not otherwise have been so. These blossom buds of course come into operation the succeeding year, while in the mean time the ring has grown up, and the circulation of the sap goes on in its usual way. If it does not grow up, but continues open from having been made too wide or from other causes, the plant will become weaker and weaker every year, and in a few years die. This stimulus, therefore, though useful with over luxuriant plants, requires to be used with caution.

Ringing after a plant is in blossom having the same tendency to stop up, and throw back into the stem and branches that part of the concocted sap or blood intended for the root, will aid in causing the blossoms to set, and in maturing the seed and enlarging the fruit of fruit trees produced the same year.

That part of a plant which has most tendency to increase its magnitude, or to produce blossoms, is the least suitable for being taken off to be used as cuttings or grafts. Thus the top shoots of heaths, and the tops of the vigorous shoots or suckers of roses, and of oranges, camellias, and all plants struck by cuttings, will not strike so readily as the less vigorous horizontal side-shoots, and especially such as are near the ground.

Although planting, transplanting, grafting, and other important operations of culture are performed with most ease, expedition, and certainty of success, at particular seasons; yet there is scarcely any season in which they may not be performed, if extraordinary care be taken in the performance, and in regulating the elements of growth, as heat, light, air, water, &c., afterwards.

These maxims refer chiefly to the green-house; and perhaps they may expand a little the ideas of such readers as may be but slightly acquainted either with practical gardening or vegetable physiology. To the gardener who is ambitious of operating on just principles, they will be of some assistance in enabling him to generalize his ideas on different parts of his art.

Sect. III. *Management of Green-house Plants in Summer.*

Whether the erection of the green-house was completed and stocked with plants in the autumn or spring, the first important operation required to be performed in their management, is the process of shifting, and this almost always takes place in the beginning of the summer quarter.

Shifting, is the transplanting of the plants from one pot to another; generally from a smaller pot to a larger, but sometimes to a smaller pot, or to one of the same size. The common object is to strengthen the roots of the plant by a supply of fresh mould; and an occasional object may be to change the manner of growth of the plant by a change of mould and a change of pot. All plants in pots require shifting occasionally, for so small a quantity of earth as a pot contains is easily exhausted; but some plants require it more frequently than others, from the nature of their roots and their habits of growth. Plants that transpire freely, as the *Geraniaceæ* and *Scitamineæ*, require more frequent renewals of mould, than such as have few pores on their surface, as the *Ericeæ*, Succulents, &c. The reason is sufficiently obvious; the former orders of plants requiring much larger supplies of water than the other.

No garden, however small, which contains a green-house, or pits and frames, can be without a shed for holding tools, pots, and other articles; and in this shed there is generally a table or bench for shifting, and for other operations with plants in pots. Preparatory to shifting, there should be a quantity of different sized pots looked out and cleaned, and sherds or pieces of broken pots prepared for cov ring the holes in their bottoms; next some loam, peat, sand, and vegetable mould, should be sifted and laid on heaps on the back part of the bench; and at hand in the shed there should be a pot or two of gravel for covering the bottoms of some pots, in addition to the sherds or crocks, some lime rubbish, props, and ties, &c.

About the middle or end of May, the operation may be commenced, by removing a few plants from one end of the stage to the shed. Then taking one of the pots, proceed to turn the plant out of it with the ball of earth entire, by turning the pot upside down, and striking the edge of the pot against the front edge of the table or bench. Examine the roots if they are healthy, much matted, or but sparingly wound round the ball of earth. If they are not healthy, the best plan will be to shake off all the earth, and cut off all the unhealthy fibres, and repot the plant in a smaller pot well drained, first by placing a sherd with the hollow side over the hole, and next by covering the bottom of the pot with an inch of gravel; and the mould used should have a little more sand added to it, than what is given to the plant in its healthy state, in order to let the water pass freely through it to the gravel.

If the ball is much matted with roots, and these are in a healthy state, it is a sure indication of the vigour of the plant, and its requiring a larger pot; and if it is desired that the plant should continue increasing in size, a larger pot may be given it. If, on the contrary, it is desired that the size of the plant should not be increased on account of want of room in the green-house, then, without cutting or breaking the roots, the ball should be broken, and the greater part of the earth shaken out from among the roots, and the plant replaced in the same pot (previously cleared and a crock put on the hole), and carefully filled in with fresh mould; lifting the pot and striking its bottom against the surface of the bench to cause the mould to settle properly about the roots, and gently thrusting a small round stick not larger than a man's little finger down among the roots for the same purpose.

If the plant is to be shifted into the same or a larger pot, this is always done with the ball nearly entire, and the web of fibres preserved untouched. Pick out from the under side of the ball the sherds or other matters used in covering the pot, and remove from its upper surface as much of the earth as is hard, sodden, mossy, or without roots. Then loosen the earth and matted roots, by gently patting the side of the ball with the hand, or by moderately pressing it between both hands. Shake off all the earth thus loosened, and having the pot in which it formerly was, cleaned, or another pot of the same size, or a larger pot, ready prepared, put in a quantity of fresh mould sufficient to raise the crown of the roots to about half an inch below the rim of the pot. On this mould set the plant, and add more earth round it, lightly beating the pot on the bench to settle it among the fibres, and using a small flat stick or spatula to press it down between the ball and the rim of the pot, taking care that no injury is done to the web of fibres. Add mould till the surface of the whole is level with the rim of the pot; and having given the last gentle beat on the bench, and seen that the stem of the plant stands upright, and at right angles to the surface of the pot, set it on a level surface beside the others which are shifted, there to be watered and stand till the earth settles.

Instead of the above mode, the former practice, and one still persisted in by some old gardeners, is, to pare off the greater part of the web of fibres with a knife, and then, without even loosening the remaining part of the ball, to set it in a large pot among fresh earth. As the mouths as absorbing pores of the fibres are chiefly at the extreme ends of the fibres, the absurdity of this practice is evident, and only the most hardy free-growing kinds of plants can withstand it. If practised on the finer heaths, proteas, and camellias, it would either kill them at once, or render them so weakly that they would not recover for several years, even if put under good management. An experienced writer on plants in pots, the late Mr. Cushing, foreman of the hot-houses at the Hammersmith nursery, states, that he believes " more plants have been destroyed by this practice, than by any other point of mismanagement whatever."

If the plant is in a healthy state, and the roots but sparingly wound round the ball of earth, the size of pot need not be increased; but the drainage and surface must be removed, and the plant replaced on a fresh drainage with a little mould, and surfaced over in the same manner. Well shaken, it may be set on the platform or floor and watered.

Green-house plants for the most part require a considerable share of pot room, especially the deciduous kinds which grow in rich soil, and have large leaves, as the *Geraniaceæ*; but great caution is necessary to avoid over-potting such kinds as grow in peat soil; these plants being always much more easily killed in large than in small pots. It is common for theoretical cultivators to observe, that the larger the mass of earth in the pot, the nearer the state of nature, the more the nourishment, and the less the influence of extreme heat or cold on the roots. But these arguments, though palpable, are found by every practical gardener to be quite fallacious. The reason seems to be, that the water which is necessarily often applied to large masses of earth in pots, when such earth is not fully occupied by roots, stagnates and rots the fibres, and the consequence is that sodden appearance, commonly called souring the soil, in which no plant however hardy will prosper.

A dozen or two of the plants being shifted, and once or twice watered with a rose or dispenser on the watering-pot, so as to settle the earth well about the roots, then begin with the first plant shifted, place it on the bench, and see if it requires pruning, tying to a prop, or otherwise dressing, before returning it to the house. Perform these things with all those shifted, and afterwards return them to the green-house, and replace them on the stage as before; unless some from disease have been so deprived of roots as to render it necessary to put them in a part of the house where they may be shaded for a few days.

In general, however, it is best in the course of taking out the plants for shifting to put to one side all diseased plants, or such as are covered with insects, or are overgrown and require cutting down, or extraordinary pruning, till the end, when they can be attended to by themselves.

In pruning, thinning, and tying up green-house plants, the greatest nicety is required, so as to preserve a symmetrical form, and yet graceful airy appearance. All cuts should be made close under an eye, or bud, or leaf, or protruding shoot, and with a sharp penknife. Elegant tapering rods painted of a light green, and small black threads, should be used as ties, neatly cut off at the knots, and not so tightly put round the shoots of the plant as to impede the progress of the sap in the bark, and cause, as often happens with common matting ties, a swelled ring above the tie.

If plants require to be washed to free them from insects or dirt, the best way is to use a soft sponge for the leaves, and a small brush for the stems and branches; and to use no liquid but pure water. Many other washes and compositions are recommended; but excepting one, which we shall afterwards mention as fit to be applied to the flues and some parts of the walls, we consider nothing so efficacious as the liberal use of clean water.

When the plants are all shifted and replaced, and those which have been deprived of most of their roots shaded, it will be necessary for a few days to watch for any changes that may take place. If the weather happens to be cloudy, it is probable nothing will occur; but if there should be much bright sun-shine, some of the plants that have had a good deal of earth taken from their roots, will probably flag or droop their leaves. When this occurs, the best mode is at once to shade the whole house during three or four hours of the middle of each day, by spreading mats over the roof. This and gentle waterings twice a day will completely re-establish the whole.

This brings us to the conclusion of the first grand summer operation, and indeed the most important of the whole year. The next is the taking out the plants, and placing them in their summer quarters.

The middle of June is the usual period for placing green-house plants in the open air; though the hardier kinds, as myrtles, diosmas, correas, and most of the green-house plants which are natives of the south of Europe, might be put out by the end of May. An old rule of gardeners was, to venture them out when the mulberry had expanded its foliage.

The object in view in placing out green-house plants, is to harden and invigorate them by exposing them to the weather, and to lessen the trouble of attending to them, by enabling them to receive the natural rains and dews, and by rendering less frequent the labour of opening and shutting the green-house sashes. In part also the object is, to get the use of the green-house for growing a few tender annuals.

The situation in which they are to be placed in the open air, should be sheltered from high winds, but not locked up from moderate breezes; it should be somewhat shaded, so as to moderate evaporation from the surface of the leaves of the plants, and which is greatly increased by the direct influence of the sun, and the greater dryness of the air with which they are now surrounded than that of the green-house: and it should have a surface impervious to worms, as, next to over-watering, nothing is more injurious to plants in pots than these vermin.

If there is no platform with all these requisites, then situations may be chosen in the north or east side of walls or hedges, or a situation may be prepared on purpose. This is to be done by surrounding and intersecting it by wattled hurdles (pannels of wicker-work generally 5 or 6 feet long and 4 or $4\frac{1}{2}$ feet high), and by levelling the ground, saturating it with lime-water, then coating it over with slacked lime and gravel, or lime and ashes, or soapers' waste, and lastly rolling these till they form a smooth compact surface. Properly executed, this sort of platform will keep out the worms for one season; but a paved surface, or setting the plants on bricks, tiles, or boards, is preferable.

It sometimes happens that there are not unsuitable places for a part of the green-house plants in walks and paved court-yards about the house; and this is not only a fortunate circumstance for the plants, but adds a peculiar and luxurious charm to the dwelling. Even lining the sides of broad gravel walks with pots of the hardier sorts has a fine effect, and they will pass their summer there very well, and not receive many worms from the well-rolled gravel.

Formerly it was the practice to plunge the green-house plants, during summer, along the borders of the kitchen garden, or in front of the shrubbery, and sometimes in groups by themselves in the flower garden. Large orange trees, myrtles, and free-flowering geraniums, produce a charming effect in this way, and may still be so treated; but the practice would be ruinous to the more tender plants; and even geraniums, and other rapid-growing spongy plants, never recover during the whole succeeding winter the check of taking them out of the ground. In fact, when geraniums are so treated, it should be with a view of never returning them to the green-house; but either of letting them remain till they are killed by frost, or of taking them up when first touched by cold, cutting off their tops, and preserving the roots in sand in a dry cellar during winter, for re-planting in spring. In this way some horticulturists have treated many of the hardier pelargoniums annually, and thus obtained a great accession of beauty to their borders and parterres.

Large orange trees sunk in the ground, so as to conceal their boxes, and then the boxes and general surface covered with turf, have a charming appearance. Those in the flower garden at Nuneham are so treated annually, and confirm our observation. In some other places, where the orange trees are planted in the soil, the entire superstructure is removed during summer, and the ground turfed over. This also is charming, and occurring unexpectedly in a fine pleasure ground or flower garden, reminds the travelled spectator of the orange groves of Genoa and Naples.

On whatever sort of surface or situation the plants are to be placed, they must not be crowded so as to exclude the air, sun, and weather from their sides and pots; nor must they, as before observed, be tied at the tops of their stems, but at a small distance above their pots. The same general principles of arrangement should be preserved, as in setting them on the green-house stage. Each genus or natural family ought to be kept together in all the groups, beds, or other masses in which the plants may be set: the tallest should be placed in the middle, and the slope to the margin or margins should be free and irregular, and not of that stiff shorn appearance before reprobated.

The plants being thus established in their summer situation, all that becomes necessary is to look over them once a day to see that none are blown over, or otherwise injured :—to observe and supply any that may want water :—to examine the ball and drainage of such as appear over-watered, and to take care that they be sparingly supplied for the time to come. In some pots worms may probably find their way; and they may either be killed where they are, by watering with lime-water, or the ball may be turned out of the pot, and the worm, which will generally be found outside of the ball, may be picked out by hand. Any vermin that may appear on the leaves should be picked off, or washed off by the syringe; but in general the small birds keep down the aphides and other insects which attack these plants during the summer. All dead leaves must be picked off as they appear; all weeds pulled out, and any holes or derangements, by watering or other causes, in the surface of the pots, filled up with fresh earth.

During dry weather, watering will have to be performed every day; the best time is the evening, and the safest plan is first to supply all the pots according to their wants, and then, if the weather is not more than usually cold, to syringe them over their tops.

Sect. IV. *Management of Green-house Plants in Autumn.*

The principal autumn operation is that of returning the plants to the green-house. This is generally done in the course of the month of September, commencing with the more tender Cape and South-sea plants, and ending with the more hardy species of the South of Europe.

Green-house plants are never shifted at this season, as it is not intended, or desirable, that they should continue growing during the autumn and winter months. Some plants, however, from disease, worms, the pot breaking, or some other accidental cause, may require to be fresh potted, and most of them will be improved by refreshing their surfaces with a little new mould, first removing the caked or mossy surface crust, and then stirring and incorporating it with what is below.

Those plants which have stood above the surface, should be gone over a week or ten days before it is intended to take them in, and all the roots which have grown through the draining-hole cut off close to it. Where many roots have protruded, this will give the plant a check, which is better received here than in the green-house; because the cool surface on which they stand will not dry up the roots like the dry boards of the stage. Plants which have been plunged should have their pots lifted up with a twist, which will break any roots that may have gone through into the soil, and then they can be cut off smoothly with a knife, and the pot replaced in its hole. Sometimes the pot breaks in this operation, in which case the plant must be carefully repotted.

The plants being thus prepared, the next thing is to attend to the green-house. Any repairs which may have been wanted will have been effected either immediately after the plants were removed, or in the course of the summer. Painting the wood-work, and white-washing the walls and flues, is generally done once in two or three years; broken glass is immediately mended; and the flues are swept once a year at least, and generally at this time. After cleaning them, they may be proved with a fire of damp straw, to ascertain if there are any cracks that will admit smoke.

Things being now ready, commence by bringing in the more tender sorts, and placing them in their situations. If the weather is still mild, they may remain there three or four days with the roof sashes and front frames off or quite open. Next, if there are any very large pots or boxes, as of oranges, camellias, &c., they, though not the next in tenderness, may be taken in, in order to get them placed while there is most room for men to move about. Then proceed with all the rest of the plants, excepting such myrtles, and South of Europe or hardy sorts, as are to be kept a month or two in a pit, to leave room in the green-house for a few Chrysanthemums and Dahlias. The plants also, by thus being placed thinner when first set in, will be less likely to lose the leaves of the young shoots by damping off. All the plants being put in, in a day or two, or as the weather dictates, put on the sashes of the roof. The front and end sashes or openings need not be shut for a week or two, unless in the case of very sharp winds; for it seldom happens that any frost occurs in September or October, and often not in November, that will injure green-house plants protected by the roof from perpendicular cold and rains.

About the beginning of November, most of the front sashes will be shut at nights, and probably a fire now and then made to dry off any damps or mouldiness which may appear on the pots; but during the day abundance of air must be given by openings here and there, in the roof, side, and ends.

If any Dahlias, Chrysanthemums, Stock July-flowers, or other choice hardy flowers, are to be prolonged in the green-house, they will now be set in, if not before; and care must be taken to do this in a perfectly dry day, otherwise the damp atmosphere that would be created by the evaporation of their leaves during the first night would be highly injurious.

In making fires to dry up damps, most people choose the evenings: this has the effect of increasing the damps by increasing the evaporation, which cannot be carried off in the night for want of a circulation of air. The best way is to make the fire early in the morning, and, as soon as the external air admits, to open the sashes in different parts of the house to promote a circulation, which will soon carry off all the damps. About midday the fire should be allowed to go out, and it need not be renewed in this way oftener than twice a week, and that only in very foggy weather. A very good plan, and one which in effect more rapidly removes damps than the use of flues, is to burn two or three pots of charcoal or coal coke in the floor of the house; it is surprising how completely this dries up the air in a short time, by that which passes through the ignited matters being completely deprived of moisture. Being also deprived of its oxygen, it must render the atmosphere of the house less wholesome for the human species; but this does not affect the plants, which in the mean time are cured of mouldiness, and the air will attain its equilibrium of salubrity in a very few hours afterwards.

The plants being now arranged as they were when originally placed in the house, all that is requisite is to attend to air, water, heat, and cleanliness.

Air must be given every day when the thermometer in the shade is above 32°. The thermometer in the green-house may, for the sake of air, fall to 45° without injury; but it is better for the plants that it should stand mornings and evenings at 48° or nearly so.

Water will be required in considerable quantities when the plants are first set in, as they are then still in a growing state; but the quantity must be gradually lessened as the days shorten and the weather becomes more foggy, moist, and cold. The greatest care must be taken not to spill any on the leaves of the plants, and so to adjust the quantity poured on the pots that as little as possible may run through them to wet the shelves, as these and the floor when wetted, greatly increase the damps by the evaporation which takes place from all moist surfaces, and from all water not frozen.

In frosty weather when the heat of the greenhouse depends entirely on the use of fire, it should seldom be raised above 43° or 45°. With the fire and sun it may reach 50°; but with fire alone 42° or 43° is even safer than 45°, which is apt to dry up the earth in the pots too rapidly, and to occasion frequent watering; and alternate watering and drying when plants are not in a growing state is ruinous to them, and kills many whose destruction is attributed directly to the frost. If green-house plants are kept very dry, and have been well hardened after they were set in by the free admission of air, they will not be killed or greatly injured by a few hours temperature of 35°, or a day or two of the temperature of 40°. We have known the earth in pots of Heaths, Camellias, and Mesembryanthemums, to be frozen for two days together and yet the plants live.

Cleanliness comprises removing mouldiness from the surface of the pots, by scraping it off, picking off all decayed or decaying leaves and flowers before they have time to drop, washing dirt off the leaves with a sponge, and burning tobacco so as to fill the house with its smoke, for destroying the aphides. The paths and shelves should be kept quite clean, but in doing so taking care not to cover the leaves of the plants with dust.

Where bulbs are grown in a green-house, they must or should be planted in the course of the autumn; and the earlier in October this operation is commenced, the better. There are scarcely any that are not best planted at this time, either of tender or hardy bulbs. The soils have been already mentioned : when planted, the pots may be set in a pit or frame kept well covered at nights to exclude the frost; and when they begin to grow they should have heat, either by being brought to the green-house, or by having the air of the pit or hot-bed warmed by flues or a lining of dung.

With some cultivators it is the practice to plant the *Ixiæ* and *Gladioli* in pots, and plunge them in the open garden along with the hardy bulbs in preparation for forcing. There they are covered with 6 or 8 inches of old tan, and when they have germinated an inch or two above the surface of the pots, they are taken up and brought into the green-house. We have already (page 117) mentioned this as the best plan with common bulbs, whether they are to be blown in pots of earth, or in glasses of water.

Sect. V. *Management of Green-house Plants during Winter.*

The winter quarter is characterized by no particular feature of management, though it is the most difficult to get over with safety.

In November, the first month of winter, the great difficulty is to prevent the fogs and damps from rotting off the young shoots which have scarcely finished growing, and from denuding large succulent-leaved plants of their foliage; all that can be done is to make fires early in the morning, every morning if requisite, but at least twice a week, and to open the sashes freely during the day. During this month and the early part of December, much more is to be dreaded from damps than from dry colds or frosts.

Towards Christmas the weather begins to grow very severe; but the atmosphere is now less charged with moisture than before. Less risk is now incurred from damps; but more to be dreaded from the alternate drying and watering of the earth in the pots. There is no mode of avoiding this entirely; but there are several points which if carefully attended to will lessen the evil in a considerable degree. The first is, not to attempt to raise the temperature of the house during night to above $42°$; if left at $45°$ in the evening, with a moderate fire at work, it will probably be at $40°$ or $38°$ early next morning, when the fire should be renewed to increase the heat a little, and then during the day, with the joint heat of the spent fire and the sun, it may be allowed to rise to $48°$.

The next thing is (where practicable) to cover
as much as possible of the roof, side, and ends,
with mats, which keeps up the temperature by
retaining the heat already there, instead of generating heat. By the use of mats the air of the
house is not nearly so much dried as by fire heat,
and consequently much less water is wanted.

The water used in the green-house during the
winter and spring months, should either have
stood in it for 20 or 24 hours, or be placed for an
hour or two over the furnace or on the flues, so
as to attain a temperature of 48°. This is a
little higher than the temperature of the air of
the house, because the earth of the pots will generally be found a little warmer than the air which
surrounds them; and it would be as injurious to
cool it down to the air of the house, as it would
be to cool the earth round the roots of plants in
the open air, down to that of the frozen atmosphere with which their tops are surrounded.

Air must be given with discretion; but still if
possible once every day, during what sunshine or
clear moments may occur from 11 to 3 o'clock.
Care must be taken that it does not enter in cur-
rents; for a strong current of cold air directed
against any one plant, or group of plants, will as
effectually produce destruction as if their roots
were to be frozen in the pot: some, to guard

against this, place a mat or netting before the openings during high or frosty winds.

All decayed leaves, mouldiness, and every thing contrary to the most fastidious cleanliness, should be removed as soon as they appear; and when unfortunately any plant dies, it should be taken out to the refuse heap, and its place divided among those adjoining: it will be to them an advantage, by giving them more air and light;—affording in this respect a parallel to the effects of the same event among our own kind.

The Chrysanthemums and other plants taken in to flower in October, will mostly be over by the end of December; they may therefore now be taken out: the Dahlias set in a dry shed or cellar from which the frost is excluded; the Chrysanthemums set in cold frames or in any warm sheltered spot; and the Stocks, or other annuals or biennials that have flowered and are done with, thrown on the compost dunghill. In the room of these may be introduced the hardier plants, kept out in pits and frames, and some or all of the bulbs. The pots should be nicely cleaned; and in order to produce a tasteful arrangement of the stage, it will at this time be desirable to reset the whole.

Some of the bulbs and some pots of mignionette should be placed at that end or part of the house where the flue enters, in order to receive a little more heat than the average temperature of the house. This will probably bring the bulbs into blossom soon after Christmas; especially some of the Crocuses, Aconites, blue Hyacinths, and Van Thol Tulips. In that situation the mignionette will flower freely, and fill the air of the house with its grateful odour.

Sect. VI. *Management of Green-house Plants in Spring*.

About the middle of February many of the green-house plants will be observed to be beginning to grow. Insects also at this time make their appearance, and some tender species that had stood the shock of winter now die off in a manner not easily accounted for.

Many plants also begin to show their blossom buds at this season; and as the days lengthen apace, and more light is supplied by nature, more heat and water may be added by man to encourage vegetation, and render the villa green-house not merely a protection for plants during winter, like that of the nursery-man, but a place for enjoying plants when they are not to be enjoyed in the open garden.

To the usual routine operations of the green-house, the spring season adds those of destroying insects, and propagating plants by seeds or cuttings or other means.

The heat of the house in February may be

raised to 48° or 50° during sunshine to encourage growth and blossoms; but need not
be higher than 44° in the evening, or 40° the following morning. Many maintain a lower temperature; but our object in the villa green-house is
to cause the plants to look well at this season.

The supplies of water may be increased, especially to plants in a growing or flowering state,
and to such as are in the warmest part of the
house. The bulbs will require it liberally till
they come into full flower, when it may be withdrawn by degrees, and as their foliage begins to
grow yellow and decay, it may be gradually left
off.

Air should be given in the middle of every
day if possible during February, and in mild
weather in March and April during the time the
sun is above the horizon, but not later. With
May the practice of leaving the front-sashes and
doors open all night may commence, but must be
conducted with caution.

Subsect. 1. *Insects and Diseases to which Green-house Plants are liable.*

The principal insects which infest green-house
plants are the aphis or green fly, the coccus
or orange-bug, and the acarus or red spider;
other insects, as flies, wasps, caterpillars, earwigs,

beetles, &c., do but little harm, or are easily taken, or scared. Snails and worms, when they find their way into the house, are more injurious.

The aphis or green fly is destroyed with ease and certainty by fumigation with tobacco. For this purpose there is a sort of refuse tobacco sold in the shops, and an appropriate bellows, with a receptacle in its nosle for the tobacco; by which means the smoke is blown out at the end of the nosle, through a dispenser or rose like that of a watering-pot. The mode commonly adopted by gardeners however, and one fully as effectual, and attended with less waste of tobacco and trouble, is as follows.

Either grow a few plants of tobacco on purpose for fumigation, or procure a quantity of the cheapest tobacco, which last will be found the strongest and best for the purpose of fumigation. Then taking what may be considered a sufficient quantity to fill the air of the house with smoke, open it out into leaves or fragments, and sprinkle it with water to prevent its burning too quick: next provide a middling-sized flower-pot, in the side of which near the bottom a hole must be pierced with a cold chisel and hammer; put a few lighted coals into the pot, and strew a little of the moistened tobacco over them; then by blowing with a pair of common bellows at the hole in the side, smoke sufficient to fill the air of the house may be procured in the space of ten or fifteen minutes. The calmest weather must be chosen for this work, otherwise there will be a waste of smoke; and it must be repeated two or more evenings in succession, if the insects are in great vigour.

The coccus or bug is never very numerous in a green-house. Washes of soap-suds and of soft soap and sulphur have been recommended for destroying it; but things must have been in a very neglected state when it becomes too numerous to be picked off by hand, or brushed off with a brush and a little warm water. From the under sides of the leaves of the orange tribe and other plants, it may also be washed off with a sponge.

The acarus, or red spider, is the most pernicious of all the insects which infest plants; but happily there is a specific antidote for this enemy, as certain in its effects as sulphur is in curing the itch, the ulcers of which in the human body are infested by another species of acarus. This is the application of water to the leaves of plants infested by this insect. In common cases water may be applied liberally from the rose of the watering-pot, or from the watering-engine or syringe; and in cases where this mode might over-water the roots, water must be applied in the form of steam, by watering the flues after they

have been heated by a brisk fire, or by raising
the temperature of the house in a cloudy day,
or in any afternoon or evening, to 50° or 60°,
and then watering all the paths and floor where
it will not injure the plants. Other modes might
be mentioned for generating steam for the destruction of these insects; but they are never so prevalent in a green-house as they are in hot-houses,
and other forcing-houses, and even in cucumber
and melon frames, and therefore the less care is
requisite to destroy them.

A mode may be mentioned for destroying these
insects without the use of water, which, though
chiefly applicable to plant-houses kept at a higher
temperature than the green-house, may sometimes
be used there, especially as it kills the aphides.
It is to wash the flues with flower of brimstone
mixed with a little white-wash. This is practised
by that curious cultivator of plants, Mr. Curator
Anderson of the Chelsea Botanic Gardens; and it
is said that half a pound of brimstone will kill all
the spiders and all other vermin of the insect kind
in a house 100 feet long and of ordinary breadth.

Raising the temperature of the house, and
watering the paths with volatile alkali, has been
tried on a small scale, and is said to produce the
same effect; but this is rather an expensive process. A similar end has been attained in the

culture of the pine apple by watering with urine ; but this leaves effects so long felt by the olfactory nerves, that it cannot be recommended for a villa green-house. Water , therefore, and chiefly applied by the syringe, may be considered as the best cure for the red spider.

A precautionary measure against this and other insects may be mentioned : it is to wash once a year, at the season when the plants are returned to the house, all the joints and crevices of the wood-work, walls, shelves, stage, and all crevices whatever, with a mixture composed of soft soap and sulphur. This is said to destroy the eggs which may be deposited in these places. We question very much if it will; but if it does not, there can be no great expense or loss of time in using it. A wash of quick-lime and water we should think likely to do as well; but in truth we do not think that in a green-house properly constructed and judiciously managed, either can ever become necessary.

Flies and wasps are chiefly injurious in the green-house during summer, when the plants are out of doors, and when the peaches, figs, and grapes are beginning to ripen. To attract the flies from peaches, place phials or bottles of water with the mouths honeyed near them; or place a few saucers of fly-water on the stage. Wasps at this season are not troublesome ; but when the grapes and late crop of figs are ripening, then the large blue fly and wasps abound. There are only two effectual ways of preserving the fruit from their attacks : the first and best is to have all the openings of the green-house covered with a fine netting, the meshes of which are not larger than three quarters of an inch. This will effectually exclude them, as no large fly or wasp can enter a space of that dimension with its wings extended, and they will rarely attempt creeping in. The second mode is to secure the bunches of grapes singly, by covering them with bags of gauze or of silver paper,—a bad plan if it can be avoided, as it lessens the flavour of the fruit. Figs may be covered individually in the same manner, and they take less injury from the process.

Earwigs and beetles are readily caught by the common beetle-trap, or by placing a glass funnel over a pot or vessel of any sort with crumbs of bread below. The beetle is less injurious where there are fruits than the earwig, which attacks the peach and fig, and is in the case of wall trees commonly caught in hollow cylinders, as reeds, or bean stalks, or the decayed flower stems of many of the *Umbelliferæ*. In these they enter for repose, and are easily shaken out and destroyed. The ant is sometimes a very troublesome insect in

the green-house, and not very readily subdued.
Where their nests cannot be dug out, or the inhabitants scalded or drowned by water, hot or
cold, we know of no mode better than wetting
the inside of a flower-pot with honey and milk, or
sugar and milk, and whitening it, covering the
hole at top with a sherd or tile.   A number of the
insects will be attracted by the odour of the honey,
and enter to collect it and settle there, when the
pot may be lifted and dipped in a pail of hot
water for a moment, and then fresh anointed and
replaced.   Fly-powder has also been tried for
destroying ants in hot-houses, and with considerable success : it does not require to be moistened,
but merely a few saucers of it placed here and
there near their haunts.   In the West Indies, ants
are poisoned with arsenic and sugar; when one
dies it is eaten by the others, who die in their
turn, and are eaten ; and thus a small dose may
destroy a large nest.

Snails sometimes find their way into the green-
house, and do a good deal of mischief before it is
well known to what or to whom to attribute it.
Their time of committing their devastations is
during night, and when day returns they bury
themselves in the soil, or settle on the dark side
of pots.   They eat the leaves of most plants, and
of some much more than others.   They are, how-

ever, easily eradicated, either in the green-house or the open air, with a little care. Either hand-pick them early in the morning, when it is hardly light; or strew over the floor of the green-house, and on the shelves of the stage, leaves of any of the cabbage tribe just beginning to decay. The snails will hardly fail to prefer the under sides of these leaves both for feeding on and as places of repose, as they can find both these at the same time, and they can be picked off every morning and the leaves replaced : or, to preserve a sightly appearance, remove the leaves during the day, and replace them at nights. In this way the market gardeners collect immense quantities of snails in the open air, and some nurserymen in their pits of half-hardy plants. Any decaying vegetable matter will do; but it is best in a state of incipient decay as being then sweeter : nothing answers better than pea-haulm.

The common earth-worm, when it infects plants in pots, may be picked out by turning out the ball; but it may also be destroyed in the pot by watering with any bitter, acrid, or acid infusion or solution, as of water of walnut leaves or tobacco, or salt-water, lime-water, or vinegar. Lime-water is the best, both as being the cheapest, and least inju-rious to the roots of the plants. Salt is equally effectual, but it destroys vegetation : it is excellent for watering garden walks, as it destroys not only vermin but moss and weeds—the two latter often very troublesome in gravel.

The common spider is sometimes abundant in green-houses carelessly managed., Common spiders are chiefly injurious to plants by soiling them with their webs, as their nests, unlike those of the red spider, are not made on the leaves, but in the angles of the walls and wood-work of the house. They are useful in catching flies to a certain extent, but form a remedy much more offensive than the disease. They are effectually subdued by the free use of water from the syringe or engine.

There are few diseases to which green-house plants are subject as such. The principal is the mildew; and occasionally the honeydew, rust and consumption make their appearance.

The mildew appears in the form of a mealy powder (supposed to be a small fungus) on the tops of the shoots of heaths, geraniums, and some other plants. A specific in general use is flower of sulphur thus applied :—Get a vessel sufficiently large to immerse the plants infested, one at a time, and fill it with clear water. Then provide a pound or more of sulphur, according to the number and size of the plants, and have ready an old powder-puff or a large dusting-brush. Then take up the plant and hold it in an inverted position, with the hand placed on the surface of the pot, so as to prevent the mould or ball from falling out : in this manner plunge the plant into the water ; and while it is wet, holding it in the same position, let an assistant dust on the sulphur with the puff or brush in such a manner that every part of the plant, including both sides of the leaves, may be perfectly covered. Then return the plants to an airy dry part of the house, the less conspicuous the better ; one dressing in this way will prove a complete remedy.

The honey-dew is a sweet clammy substance, which coagulates on the leaves of plants, chiefly during summer, but on some green-house plants, as the *Rosa semperflorens*, Myrtle, and Orange, during winter and spring. Some regard it as the dung of aphides, and it is certainly often found on the rose along with these insects ; but it is also found on the orange alone, and which the aphides seldom inhabit ; and thus it appears to be an exudation in some cases : whatever may be its nature, it appears to be more offensive to the eye than injurious to the plant. We know of no mode of preventing it : washing it off with a sponge and water from time to time is a palliative unattended with injury to the plant.

The blight is a term in common use, and is applied both to animals and plants when they

have received any disease or injury not otherwise
to be accounted for. It is supposed to be produced by east winds, extreme heats, thunder and
lightning, &c. As far as the disease has been
examined, it seems to be a minute fungus growing on the plant, and of course nourished by its
destruction. Salt water has been found to destroy it in some cases, but rarely till the injury
was already so great that the diseased plant could
not recover. In gardens the same process as for
the mildew has been tried with success, and this
is all that we can recommend. Fortunately it is
not very common in green-houses.

Various insects and diseases attack the peach
tree, as a black fly, the thrips, a very small insect
like the aphis, the acarus, a caterpillar, the curl,
mildew, blight, and honeydew. The aphis
and thrips, and also the black fly, or any other
of the fly kind, will be effectually destroyed by
tobacco smoke;—the acarus, by abundant watering over the leaves : the caterpillars are few, and
to be picked off : the curl is occasioned by minute
insects, their kind is not correctly ascertained, but
it is prevented from increasing, by the use of sulphur and watering, as are the blight and honey-
dew. Some use the precaution of washing the
shoots of peach trees and vines with soft soap and
sulphur after they are pruned ; but this we consider quite unnecessary.

The vine is attacked by the red spider, for which the remedy is obvious; the coccus sometimes appears on the shoots; and these are to be brushed off: but rarely any other insect appears of an injurious nature.

The bleeding of the vine may be treated as a disease. It takes place when pruning has been performed too early or too late : it may be checked by pouring abundance of cold water on the roots; but it is questionable whether this will not do more injury to the plant than the disease. Searing the wounds has been recommended, and also tying on a piece of bladder, thin lead, &c. Our opinion is, that the bleeding will do very little harm, and that it may be left to take its course.

What is called consumption in plants, is a disease which gradually and sometimes quickly induces the decline and decay of the plant till at last it dies. The causes which produce it can seldom be correctly ascertained; sometimes it appears to proceed from improper soil, climate, or culture; at other times it comes on under the most favourable circumstances, as far as appearance goes. Whatever may be its nature, there is no remedy; and all that the gardener can do, if he discovers it before it is too late, is to take off a few cuttings, or adopt the means peculiar to the plant for procuring a successor.

Subsect. 2. *Propagation of Green-house Plants by Seeds.*

The beginning of February is considered the best season for sowing the seeds of most green-house plants, because this affords them all the summer season to acquire a size for potting off towards autumn, and being hardened a little before winter. Heaths however are an exception; for though they will vegetate very well at this season, yet if they can be sown the preceding September after they are just ripened and gathered, they will acquire sufficient strength to stand through the winter, and be fit for potting off with other seeds sown in February. But if heath seeds are sown in spring, they should be sown very thin, so that the plants may be able to stand in the seed pots all the succeeding winter without damping off; for sown in the usual way in spring, they are not fit for potting off till it is too late in autumn to attempt it, and consequently they are left in a crowded state in their seed pots during winter, when they almost inevitably damp off for want of air round the plants. Sown in autumn, they are fit to be potted off in the July or August following, and hardened in frames before the season arrives of setting them in the green-house.

Seeds are sown in the same sorts of soil which

the parent plants thrive best in: hence, where there is a variety of seeds to be sown, a variety of soils must be prepared. Loam, peat, and sand, however, are the simples which will supply all requisite composts. The size of the pots may be from 4 to 8 inches across, as it is not desirable to sow seeds in very small pots on account of the rapid drying cf the soil. These pots should be well drained by first placing a crock or sherd, or oyster-shell with the concave side undermost, on the hole, then covering it with an inch or more of gravel or of broken sherds, and over that laying some peat siftings or fibrous vegetable matter obtained in sifting soils, to prevent the soil from washing down and mixing with the drainage, and so rendering the latter inefficient. On this bottoming, as it is technically called, place the mould proper for the seeds to be sown: press it down gently, so as it may stand at half an inch under the rim of the pot. Then lay on a quarter of an inch of the same mould sifted extra fine on purpose, and spread it quite level. On this sow the seeds as evenly as possible, and cover with an eighth or a quarter of an inch according to the size of the seeds. Then press this covering gently and evenly down on the seeds with the hand, leaving the surface quite smooth and level, and the operation is finished.

If the seeds are large, such as nuts or stones, then the pots must not be filled so full by half an inch or an inch, and the nut or stone when planted must be pressed into the soil, and the soil over it pressed well down on it, so as it may be compactly inclosed in earth.

There is not an operation among all the processes of culture more essential to growth, than that of pressing the soil to roots, seeds or cuttings, newly inserted in it. Even the common cabbage will not grow if this point be neglected. This the sagacious Cobbett adverts to in a strong manner in his book on American Horticulture, justly observing, that it is not sufficient in dibbling-in cabbages to make them firm, but they must be made firm at the extreme point of the root. The same thing may be said of all descriptions of cuttings, which, whether of heaths or of vines or gooseberries, must be made quite firm at their lower extremities. We have seen many crops, both of the field and the garden, fail for want of pressure on the seeds; and the late Mr. Cushing, who was one of our best exotic gardeners in his day, observes of seeds in pots, that he has "witnessed in many of the nurseries, as well as gardens, seeds sown, and left not only without being pressed in, but almost uncovered; the evil tendency of which in places that should set the example, at least in sowing of seeds, must be evident to the commonest observer." (*Exotic Gardener*, p. 5.)

If the parcels of seeds are small, two or three may be sown in distinct patches in the same pot. These patches may either be in the form of bars or slips across the pot, each being distinguished by its name written on a small naming-stick; or they may be in the form of triangles with their apexes pointing to the centre of the pot; and round this centre may be placed the different naming-sticks, each pointing to its triangle, like little tombstones to their graves. A cross section of an orange will convey a tolerable idea of the appearance of the surface of such a pot: the line round each clove represents the triangles; and the centre where they all originate the point where the naming-sticks are placed.

The sowing being finished, the pots must be gently watered with the finest dispenser that can be made. Repeat this at intervals of a few minutes till the mould becomes sufficiently moist for vegetation; then set the pots in a dry airy part of the green-house where they will not obtrude themselves on the eye and interfere with the general effect: or, if there is no room in the green-house, set them on a hot-bed already spent, but where, by the aid of what heat remains in the dung, covering well at night, and the sun during day, they may enjoy about the same temperature which they would have done in the green-house. The very choice sorts, and most of the heaths, should be covered with glasses fitted to the inside of the rim of the pot; and these should be kept on, excepting while water is given, till the seeds begin to come up, when they may be taken off part of the day, and in a few weeks entirely. All the culture these seeds will require till they come up, will be gentle watering, so as always to keep the surface of the soil moist, and weeding. The weeds should be pulled when in their seed leaf, otherwise they are apt to bring up the seeds with their roots, and thus, if they are beginning to germinate, destroy them.

As the weather becomes warm, and the sun shines bright, the pots will require to be shaded for an hour or two in the middle of every day. This may either be done by a mat if they are in a frame, or by sheets of gray paper if they are in the green-house. When the plants begin to come up, this shading must be increased. In two months most of the seeds, excepting stones and nuts, will have vegetated; some of these require to stand a year before they come up, and they must be kept during this period in the same temperature as green-house plants.

Watering, weeding, and shading having been attended to till the middle of June, all those pots whose seeds have come up may be removed to the open air, and placed in the shade of a hedge or high wall, or of wicker hurdles. Here being kept moist and weeded, and worms, slugs, and other vermin kept from annoying them, they will be found fit to pot off by the middle or end of July.

It is very desirable to begin to pot off as early as possible; because while the plants are young and small their roots have not had time to mat together as they do afterwards, and they therefore not only receive less injury by the operation, but the early period at which this is done gives them a longer season to establish themselves before being stopped in their growth by winter.

In potting, be careful to match the pot to the size of the plant, and its rapid or slow growth. The Geranium tribe will require the largest pots; but these need not exceed the size called sixties, or half-penny pots, and which are only about two inches in diameter; Heaths, Diosmas, and many such small plants, are better potted into thimble pots which are under an inch and a half in diameter. In potting, let the mould be quite dry, and gently press it to the roots with the fingers; then water with the dispenser once or twice till the earth is well settled; after this set the pots on a cool frame or a pavement, or on a bed of ashes, or other bottom impervious to worms. Keep the lights close on night and day for a few days, opening only to water, and shade during all the time the sun shines brightly on the frame. In a fortnight begin to harden them by degrees, first taking off the lights at night, and shutting them and keeping them shaded during the day; next taking them off the greater part of the day, and lastly entirely; only replacing them during very heavy rains and storms.

Thus treated, some of them, as the Geraniums, may be set along with the green-house plants in August, and others allowed to remain in the frames till they are removed to the green-house with the main stock in September.

The pots of seeds which have not germinated must be kept during the winter either in a concealed part of the green-house or in a frame or pit of similar temperature, and when spring comes treated in all respects like the seeds which vegetate the same season in which they are sown.

It may be observed here, that it is a safe plan with new seeds from the Cape, Australasia, or South America, to place them when sown on a heat rather higher than that of the green-house; for though the plants of these and other latitudes

when grown will thrive very well in our green-
houses, yet the seeds require a greater degree of
heat than is kept in such structures, to excite them
to vegetate.

Subsect. 3. *Propagation of Green-house Plants by Cuttings.*

Cuttings in general should be put in at as early
a period of the season as possible, for the same
reasons as are given for the early sowing of seeds,
viz. that the plants may be rooted, transplanted,
and hardened before winter.

Cuttings of most kinds of green-house plants
are best put in in spring; though some prefer the
autumn for coriaceous evergreen-leaved plants, as
the Camellia and Orange. In the latter case the
cuttings remain in a nearly dormant state during
October, November, and December, and are then
supposed to have formed a callosity at their lower
ends, ready to emit roots when they are put into
bottom heat in January.

Most plants succeed best when the cuttings
are taken off in young succulent shoots; and
hence it is necessary that the plant to be propagated be in a growing state. For this purpose,
it is the practice with nurserymen and others who
propagate extensively, to place specimens of the

plants considered more rare or difficult to root in
the stove early in January, that they may protrude
shoots ready to be taken off in February. With
the more common kinds, or any not much in demand or readily struck, they wait till they are sufficiently grown with the usual temperature of the
green-house. As the villa winter garden will
generally be kept at a higher temperature than
the green-houses of nursery-men, where the object
is to preserve the plants, more than to produce an
effect; and as it is only now and then that a few
cuttings can be wanted for keeping up the stock,
abundance of young wood will be found without
recourse to forcing.

Though most plants strike readily in young
wood, and with leaves on the upper part of the
cutting, yet there are some which root very well
in old wood, or wood of one or two years' growth,
as the orange, the leaves being retained; and
others of the deciduous kind which root freely in
wood, one, two, or three years old, without any
leaves. This however applies only to the deciduous sorts, as *Fuchsia*, *Aloysia*, *&c.* Some plants
root most readily when the cutting is just beginning to ripen its wood, and others when the wood
is ripened. No general rules are in this respect
universally applicable, and therefore the unexpe-
rienced cultivator may consult our catalogue, in

P

which the most successful modes are noticed; and with quite new plants, of which there may be said to be no experience extant, he must exercise his own sagacity, and find out what is best, by observation and experience.

For most plants that are propagated by cuttings, the same soil and climate in which the parent plant thrives are allotted; but some which are very difficult to root are struck in pure sand; and as to climate, most species strike the readier in one somewhat more moist and warm than that which the parent plant inhabits. Almost all cuttings strike the better for being covered with a bell-glass, which, by stagnating a portion of air round the cutting, renders all aërial changes less felt; maintains a congenial moisture in the air, which though generally injurious to animals is yet highly favourable to vegetables; lessens the stimulus of light; and tends, being in the shade, to maintain an equable temperature.

The operation of taking off, preparing, and planting cuttings, cannot be commenced, carried on, and completed in a day or two, like that of shifting or potting plants. All sorts do not come into a fit state for having cuttings taken off at the same time: some will be ready earlier, and some later. Whenever a plant appears to be in a proper condition for the purpose, it should be proceeded with, and the others left till they are ready.

Pots, soils, sherds, gravel, and siftings of peat or other mould, are required to be at hand for the operation of planting cuttings, as for planting seeds. Having fixed on the kind or kinds to be propagated, first consider whether they will root best in loam, peat, or sand. Drain the pots as for sowing seeds, and fill them brim full with the proper compost, shaking the pot and pressing down the compost with the hand : then having taken off the points of the shoots of an inch or more in length according to the kinds, proceed to cut off their lower leaves with a sharp penknife to half the length of the cutting, and sometimes more; and lastly, cut it clean across at a joint or bud, placing it on the nail, or on a thumb-piece for that purpose, as is done in cutting the nib of a pen.

When the cuttings are to be planted in peat or sand, the pots being filled will have to be previously well watered so as to render the whole of the soil moist, otherwise it cannot be sufficiently tightened to the base of the cutting ;—a point, as we have observed, more than once, of the most essential consequence. Loam will generally be found sufficiently moist for that purpose without watering. Not more than one species should be put in a pot, on account of the very different periods different plants take to root. In the case of very scarce or delicate sorts, it is common to put only one cutting in the centre of a small pot, which renders them less liable to be injured by damp, and early transplantation is not wanted.

In planting, use a small dibble to make a hole for the cutting, and press the earth or sand tightly to its lower end. In general they should not be inserted so thick as that the leaves will touch each other, and sufficiently distant from the edges of the pot to admit of their being covered with a bell-glass. As soon as each pot is planted, and the surface of the mould made level and firm, give them a gentle watering to settle them: leave them to soak for a quarter of an hour, and then cover them with a bell-glass, which should be pressed sufficiently tight to exclude the outward air. The pots should now be plunged in a gentle hot-bed, or placed in a shady part of the pit of a hot-house.

Being plunged in a pit or hot-bed, the culture for some weeks consists solely in taking off the glasses every morning, and wiping their insides perfectly dry with a cloth, and replacing them; removing at the same time any cuttings or leaves which happen to damp off. When the sun shines unclouded, they must be shaded for a few hours daily, moderately, by a mat if in a hot-bed, or by sheets of paper if in a hot-house pit; but care must be taken not to leave on the shade too late in the afternoon, as the soft tender state of the cuttings renders them very liable to damp off by over shading.

In ten days or a fortnight the free-rooting kinds, as the *Geraniaceæ*, will be indicating a state of incipient growth. When this is noticed, it will be necessary to give them a little air by taking the glasses off every morning when the sun is quite receded from them, and putting them on again early in the following morning. By this means they will in the course of a few weeks become so hardened, as to be able to bear the full power of the sun without the glass; when the shading and glasses may be laid entirely aside. If any of them should droop their heads when this operation is first performed, it is proper to refrain from moving the glasses until they have gained more strength. The shading is also to be decreased by degrees, but not so much as to be entirely done away while the glasses are in use.

The operation of planting cuttings, and their routine culture in this way, may go on from the beginning of February to the beginning of July; after which only the most hardy and easily-rooted sorts should be attempted, on account of the little chance of getting them potted off and hardened before winter. Geraniums, however, may be propagated by cuttings till September, by placing only one cutting in a small pot, and then plunging the pot in heat : the cutting in this way becomes a neat little plant in the course of a month, and will stand the winter in the same pot.

The most difficult green-house plants to strike are heaths, the particular mode of operating with which we have already given (page 59). These and almost all the woody shrub-like plants, as the *Proteaceæ, Myrteaceæ, Diosmaceæ, &c.*, propagate most readily during the months of May and June, when the wood is in a growing state. Any one may convince himself of this fact, by putting in a few cuttings of the growing wood of a common myrtle in May, and again a few of the ripened young wood of the same myrtle in August. The former will root almost immediately, and make neat little plants before winter ; whereas the latter will only form callosities on their lower ends, and must be kept in the pot till spring, when they will throw out roots from these callosities and begin growing.

In the beginning of May, some of the cuttings put in in February will be fit to pot off. This must be done with the same care as in potting off seedlings. After being watered, they should be

set in a close frame with a little heat, or, if there
are only a few of them, under a hand-glass in the
green-house. The frame or hand-glass should be
kept close over them for a day or two, till they
show signs of having rooted into the soil : then
they may have air daily, beginning by a sparing
admission, and shading during bright sunshine.
As the plants appear to be hardened, increase the
air till the sashes or hand-glasses are taken off
altogether. After this, watering, weeding, picking off dead leaves, attending to the neatness
of the surface of the pots, and to the exclusion of
vermin, are all that is necessary. The plants may
either be set among the others on the prepared
platforms, or, what is better, by themselves under
the shade of a wall or hedge. Some of the *Geraniaceæ* will produce fine strong flowers in
August, and most likely ripen a few seeds.

Some of the latest-planted cuttings, as those of
many of the Proteas, Banksias, and other hard-
wooded Australasian plants, as well as Camellias,
will not be fit to pot off the same season. These
may have their glasses taken off (as leathery-
leaved plants require them much less than those
with tender foliage), and be set in an airy, rather
shady part of the green-house, not exposed to
view so as to injure the general effect. In the
succeeding February they may be placed on a

little heat, when they will push freely. In the Nurseries, cuttings of Camellias and of the *Citrus* tribe are generally not taken off till the shoots have finished their growth, the leaves attained their full size, and the bark at the lower end of the shoot begun to get brown. They are then planted in broad pots of loam, and kept without being covered with glasses in an airy part of the green-house all winter, and in February put on heat. When growing shoots are used, they are planted in sand covered with a bell, and immediately put on heat: in this way a number of them strike root; but if young shoots are planted in loam, whether put in heat or not, and whether with or without a bell-glass, they generally damp off. But on the Camellia and Citrus we have already treated (pages 39 and 81).

Subsect. 4. *Propagation of Green-house Plants by Laying, Inarching, Budding, and Grafting.*

A few green-house plants are propagated by laying; and this mode is also sometimes adopted for such as strike freely enough by cuttings to get larger plants sooner, or for the sake of running less risk of failure where there are few parent plants, and perhaps little skill in the operator. Inarching, budding, and grafting, are practised with a few species, and very commonly with the fine varieties of Orange and Camellia. Some of the *Myrteaceæ* and *Proteaceæ* are also so multiplied. On the continent most of the greenhouse plants which are difficult to root, are propagated by layers, and by the Chinese mode of ringing, and inclosing the ringed part of the stem in earth. In China they knew no other mode than laying, before grafting was introduced by Europeans. It is remarkable that the Otaheitans, who know neither grafting nor laying, are yet acquainted with the mode of propagating by leaves, a mode which we shall notice more particularly at the end of this section, and which is also a natural mode like laying.

Laying is a very natural mode of propagating plants of the woody kind, as we see the dependent branches of several species of trees resting on the soil and rooting into it. It is seldom that trees have sufficient room, and are so far left to themselves as to have branches depending on the ground; but where the lime, holly, hazel, thorn, sycamore, &c., have attained a great age in some old woods, rooted branches may be found round their trunks. There is a fine example of the lime tree rooting in this way at Knowle in Kent, and we have seen elsewhere even the spruce fir and cedar-larch so rooted.

The improvement which man has made on the natural mode of layering, is to bury the shoot deeper in the soil, by which means the stimuli of moisture, the very·considerable one of pressure, and the necessary condition of the exclusion of air, are increased. Another improvement is that of wounding that part of the shoot which is buried, by which the tendency to put out roots is greatly increased, partly by the quantity of nourishment derived from the stock or parent being diminished, and partly from the returning sap being impeded and stagnated in the layer, at the lower end of which it first produces a callus, and then roots.

There are different ways of wounding the shoot to be layed : a very common mode is giving it such a twist as shall fracture longitudinally both the bark and wood ; another mode is by thrusting a knife through the shoot, withdrawing it, and then inserting a bit of wood, slate, stone, or a nail, in its stead . some twist a wire tightly round the part from which the roots are to protrude ; others pierce it with an awl in two or three places : the most common mode is to make an incision half way through the shoot, and return it longitudinally upwards or towards its extremity, so as to separate about half an inch of half of the shoot ; this half inch is technically called the tongue, and the process tonguing. About one third of this

separated part or scalp (being what is generally
fractured by the operation) should be cut off in a
transverse direction, and directly across a bud or
joint, as in cutting the ends of cuttings. This is
an essential part of the operation, neglected by
many gardeners who have never considered the
mode in which cuttings throw out roots. The
last mode of wounding or preparing a layer which
we shall mention, is that of taking off a ring of
bark half an inch or more in width, below the
part where it is intended the roots should originate. This ring should always be made right
across a bud, removing one half of it : its use is to
impede the returning sap, which it does effectually, and young roots soon protrude from the
callosity formed on the upper edge of the bark.

Whatever mode be adopted, three things are
essential : first, to bury the shoot at that part
where it has undergone the process intended to
promote the protrusion of roots, an inch or two in
the soil ; secondly, to bind up the point of the
shoot, so as it come out of the earth in as near a
perpendicular attitude as possible ; and thirdly, to
press the earth very firmly to the part intended to
root. A further precaution is to peg down the
layer with a hooked stick, to prevent it from
rising or being disturbed ; but with green-house
plants this is seldom necessary, more especially

if the soil is loamy and firmly pressed to the layer. Any person may prove the value of pressure, by laying two shoots of the same species in similar soil, and at a similar depth, pressing neither, but well watering each, and then, after the water has subsided, laying three or four bricks above one another over the one of the layers. We have seen this tried with the vine, and found the difference of effect truly astonishing.

In laying plants in a private green-house, a very convenient mode is, to place a small pot on the surface of the larger one in which the plant grows, and lay down the shoot in that, fixing the pot with a pin so that it may not be disturbed. Sometimes it answers to. draw the shoot to be layered through the bottom of the pot, and midway up the pot, to make the incision; but this though in common use in laying vines, and some creepers as the *Passifloreæ*, is not generally applicable to green-house plants, owing to the shortness of their shoots. In nursery gardens, where laying green-house plants, as the Camellia, Ilex, and some other difficult sorts, is practised on an extended scale, the whole plant is layed down on its side in a pit or house on purpose, and every part of it is turned into progeny.

Layers are layed into the same soil as that in which the parent plant thrives the best; but with the addition of a little more sand to lessen the risk of soddening the soil by too much water. In general the layers should be formed of the young tender shoots of the present year, the soft bark of which will sooner form a callosity than that of any of the preceding year's growth. Next to these, the shoots of the preceding year are to be preferred.

It is particularly necessary to be observed, whether the plant is of a brittle nature or not: for if it is, it will require very gentle treatment, and the points of the layers instead of being turned perpendicularly upwards, as should generally be done, must be left in an inclined position until they commence growing, when their tops will recur to the natural position of all plants.

Where entire plants are layed down to produce layers, they are called stools; and the main root remains there as a stool for several years, and affords an annual or biennial supply of shoots for laying down. In this case, care must be taken not to cover any part of the stool or of the root ends of the shoots with earth, beyond what is necessary to cause them to strike root, as that is very apt to rot them, and prevent them sending up shoots for another year.

The Chinese mode of propagating trees may next be mentioned, as nearly allied to laying.

The objects which they have in view are of two kinds; one is in the case of rare trees, merely to propagate them; and the other in the case of common sorts, to produce dwarfs, or miniature timber and fruit trees. Thus they have in small pots, branches of firs, pines, and cedars, bearing cones; and of oaks, chesnuts, and beech, bearing acorns and must. For propagation they make choice of a shoot of from one to two or three years growth; and at the distance of a foot or more from its extremity, cut off a ring of bark an inch or two long; over the decorticated part, and extending two or three inches above and below it, they fix a ball of prepared loamy earth, being loam mixed with rich desiccated night soil or other manure and a little sand; this they cover with moss, or several folds of cloth, and the whole is firmly tied to the shoot with cords or shreds of the bark of elm or lime. A strong stick is next inserted in the ground, and to this the shoot and ball of enveloped soil is tied to keep the whole secure; and lastly, a porous earthenware vessel is supported by the stick, so as to hang over the ball; and this being filled with water, a few drops per day ooze through the sides of the vessel, collect on its bottom, and then drop on the ball. When the water gets low a fresh supply is added, and this is all the culture required till the roots are observed, on unfolding the envelope, to have pierced the mould, when the shoot is cut off with the ball, the moss or cloth taken off, and the plant with the ball undisturbed planted in a pot or in the open soil according to the kind. Rooting generally takes place in six months, but with some species a year is required.

The next mode, which is for the purpose of procuring dwarf forest trees, is to fix on a handsome branch of a tree in full bearing of fruit, cones, or seeds; then at a convenient part free from side branches, where it may be from 1 to 2 inches in diameter, cut off a ring of bark, and proceed exactly as in the other case. This branch will sometimes, as in the case of the resinous tribe, require two or three years to root; but that being effected, the branch is cut off and potted, and is considered a great curiosity. Such dwarfs seldom increase much in size, but go on bearing a few years and then decline and die.

The continental gardeners propagate a number of rose plants on the same principle, and evidently in imitation of the Chinese. They ring a shoot, and instead of tying a ball of earth round it, they have little tin vessels with a slit on one side, extending from the brim to a hole in the lower part of the side or the bottom. The shoot being prepared, the vessel is stretched so as to open the slit, in which the shoot is passed, till it comes to the hole where it remains, and the slit closes again by its own elasticity. The hole is then stopped, but not so as to entirely prevent water from escaping, and the pot or can being filled with the proper mould and watered, the operation is nearly completed. All that is done afterwards is in some cases to tie the tin case and shoot to a supporting rod, and sometimes to cover the former with moss or rags. Some use earthenware vessels made in imitation of the tin ones; others, little wooden boxes, cocoa-nut shells, and we have even seen necks of bottles so applied. The principal part of the culture is not to neglect regular supplies of water.

But these are clumsy modes, tedious, troublesome, and expensive, and therefore unworthy the imitation of British gardeners.

From laying we proceed to inarching; which may be called laying into wood, instead of laying into earth. It is a species of grafting, but differs from it in these particulars; that whereas in grafting the scion is at once totally separated from its parent plant, and the head of the stock is cut clear off before the splicing or fitting together takes place; here, on the contrary, neither the scion is separated from its parent, nor (in general) the head of the stock cut away, until the union

becomes so far complete that the first is unnecessary, and the latter injurious. It is in consequence
much preferable to the common mode of grafting,
for all delicate and difficult plants, and especially
for evergreens. It is generally practised as the
best means of multiplying all the fine double
varieties of camellia, and plants of similar habits ;
because their strong leathery leaves, if only for a
few days deprived of their regular support, by
being cut clean off from the mother stock, if not
covered very closely with a glass will be certain
to wither and fall off, after which there is very
little chance of the scions completing a union with
the stock.

It is an important point in grafting of any kind,
or in budding, to ascertain what sort of stocks
may be used, and what may be the probable influence of the stock on the scion. The general
and a safe rule is, to choose a stock from some of
the coarser free-growing varieties or species of
the same genus; and thus the common red
camellia is chosen as a stock for most of the other
sorts ; the common citron for the different sorts of
orange; and so on. But sometimes the same
genus does not afford a coarser-growing species,
or perhaps there is no other species of that genus.
What is to be done for a stock in this case? All
we can say is, Choose some free-growing plant of

the same natural tribe or order, and as nearly allied to the genus as possible. This will not always answer, but it very often will, and at any rate it is the means by which sagacity must discover experimentally what will finally succeed. In general, all plants of the same natural family will breed together; that is, produce mules, if the operation of impregnation be carefully effected by art, and will also bud and graft together. There are, however, numerous exceptions: thus, no one has ever (as far as we have ascertained) been able to produce a mule between the black and red currant, often as it has been tried; nor will a pear grafted on apple, or the contrary, last above a year or two.

With respect to the effect of the stock on the scion, that is as difficult to determine *a priori* as what plants will succeed grafted on each other. Arguing from what takes place in the case of fruit trees, we would say that vigorous-growing trees will be dwarfed and thrown into a blossoming state, by being grafted on low-rooted kinds with few ramose and many fibrous roots; as is the apple by being grafted on the paradise variety, and the pear on the quince; and in like manner, that less robust kinds would have their vigour and duration somewhat increased by being grafted on free-growing branchy trees, as are the finer varieties of apple and pear by being grafted on the crab and wilding pear. Whether this universally holds, or not, we cannot determine; but in the present state of our knowledge it is safe to act as if we thought so. As to the influence of the stock in changing the qualities of fruits, or the colours or fragrance of leaves or flowers, that is very uncertain. Quinces have been said to render some sorts of pears gritty; and crab stocks to lend acidity to some of the milder apples; but these things are denied by some, and doubted by most who have considered the subject. One curious fact may be mentioned, which is, that a variegated jasmine grafted on one not variegated, will transfer the variegation to the whole plant, and even to suckers rising from its roots. There is a proof of this in the botanic garden at Chelsea; and the same thing is said to take place with a variegated privet on a plain one, and with some sorts of *Phillyrea*.

The manipulation of inarching is thus given by Cushing, as applicable to exotics in stoves or green-houses: " Having provided a stock nearly of the same diameter as the shoot which is intended for inarching, cut a thin slip from two to three inches long, and about one-third or something better of the whole thickness, smoothly off from each of them, in the clearest part of the stem, with a sharp knife; the bark of each must then be

fitted together exactly, and tied perfectly tight
with good matting; they must be clayed in the
same manner as grafts; and as being within doors
in a warm house will occasion the clay to become
over-dry, and in consequence liable to crust, they
should (at least in hot weather) receive two or
three times a week some water from the rose of
a watering-pot, or by means of a syringe, to preserve it in a moist proper state, observing to do
it in the evening, lest the leaves should get
scorched by the rays of the sun: a little moss
tied neatly round each ball of clay will prevent
the watering from being so frequently necessary,
and is therefore a very desirable addition." Eight
or ten weeks will in general be found sufficient
time for them to unite; at all events, by that time
they may be partially separated from the parent
plant, by cutting the inarched shoots more than
half-way through; and if on trial they are found
to be united, and bear that operation well, they
may in a few days afterwards be entirely cut off
and placed in a shady part of the house, where
they must be kept moderately syringed as before,
and some additional shade given according to the
state of the weather for two or three weeks, during
which time they may be untied, and the top of
the stock cut off in a neat manner; and also any
unnecessary part of the lower end of the scion

that may remain. Then apply a little clay to facilitate the healing of the wounds, which will take place in a few weeks, when the plant may be considered as established and fit to be placed among the general collection.

There is a method of propagating some green-house plants, especially camellias and oranges, which Cushing describes as mid-way between inarching and grafting. By this mode the top of the stock is left on as in inarching, but the scion is cut off as in grafting : it is chiefly applicable to such plants as can be covered with bell-glasses.

A seedling lemon or orange of a year old being procured as a stock, "choose a scion of similar diameter, and cut the lower end of it in a sloping direction as for the common whip graft. Then, without taking off the head of the stock, cut from the clearest part of its stem an equal splice, as smoothly as possible, so as to be fit to receive the scion : let neither stock nor scion be tongued, but apply the scion to the stock in a neat manner, so that their barks on both edges and below may join, and then tie them in a firm manner with matting, and clay them as in grafting. Then cover with a tall glass or receiver (technically a cap), and plunge in a moderate bottom heat, and shade and treat in all respects as cuttings." In six weeks the scion will have begun to grow, and the head of the stock may be cut neatly off, and the clay and ties removed by degrees.

Grafting is occasionally practised with greenhouse plants, but it is not in very general use. There are many varieties of grafting, but the most general is the whip method commonly adopted by the nurserymen in grafting fruit trees. It can hardly be considered necessary to describe the mode of performing the operation, it being so generally known; and it would be of little advantage to describe one method, unless others were also added. Those amateur cultivators who wish to acquire the practice of it, will learn more from observing an expert gardener perform the operation a few times, than from several pages of description.

The success of grafting, whatever kind be employed, depends chiefly on the exact union of the inner bark of the scion with the inner bark of the stock, by which means the sap is enabled to ascend from the soft wood or alburnum of the latter through that of the former, and to return in like manner through their united barks. The other requisites to success are of inferior consequence, but still deserve attention : they are, that the sap of both scion and stock must be in motion, and that nearly in the same degree in each; and that the operation be performed with celerity.

In grafting, it has been observed that the nearer the graft is made to the root of the stock, so much greater is the probability of success. Hence green-house and other rare plants are sometimes grafted very low, and sometimes even on the roots. When low, in addition to claying, the graft is earthed up to the upper edge of the clay, or the lowest bud of the scion, and this is found of great use by preserving an equable degree of moisture, and preventing much waste by transpiration. Another advantage of low grafting in the green-house is, that the plant can be covered with a cap- or bell-glass, which is always of advantage.

As we have described a mode of inarching nearly allied to grafting, as practised by Mr. Cushing, so we shall here describe a mode of grafting nearly allied to budding as practised by Mr. Nairn, a very ingenious man and an excellent practical gardener. This mode is far easier than either inarching or common grafting, and therefore deserves the particular attention of the amateur. Mr. Nairn applied it to orange trees, and gives the following directions:—" Let the operator select as many orange or lemon stocks as he wishes to work, and place them on a moderate hot-bed for a fortnight, by which time the sap will have risen sufficiently to move the bark: the stocks must then be cut off about two inches above the surface of the pot, and an incision made with a sharp knife similar to what is done for budding, separating the bark from the wood on each side. Let the scion be cut thin in a sloping direction, and thrust between the bark and wood, and then bound tight with woollen yarn; but very great care must be taken in binding, to prevent the bark from slipping round the stock, which without attention it is very apt to do. After it is properly and neatly bound, put a little loam or clay close round the stock to the surface of the pot, then cover with a cap-glass. This cap-glass or receiver must be of a form bellied out a little towards the top, and with the top more acute than is usual, in order that the condensed damp that will collect on it may run down its sides and not drop down from the roof on the scion. The glass must be pressed firmly into the mould, to prevent the air or steam from getting to the plant, and must not be taken off unless any of the leaves are found damping off, and then only to remove those; when it must be immediately replaced and made air-tight. The operation of grafting and putting on the glasses completed, plunge the pots on a hot-bed with a brisk heat, and in about six weeks the glasses may be taken off and the clay and binding removed; but it will be necessary to bind on a little damp moss in lieu of the clay, and to keep the glasses on in the heat of the day, taking them off at night. In about three weeks or a month, they will be in a fit state to be taken into the green-house, when they will be found one of the greatest ornaments it can receive." (*Hort. Trans.* vol. iii.)

Mr. Nairn has in this way grafted scions with blossom buds, and had ripe fruit the same year of grafting :—he has gone even so far as to graft with both fruit and flowers on the scion. When this is to be done, he recommends the mandarin as the best to commence with, as the fruit is more firmly fixed than that of any of the other sorts : he says he has had seven of these oranges on a plant in a pot, commonly called a small sixty, which he justly observes is one of the most curious and handsome ornaments that the green-house can receive.

One or two green-house plants are propagated by budding, and more might be so increased, especially if the scallope or French mode of budding were adopted. The common mode of budding, by inserting a bud with a portion of bark attached, under a piece of bark raised up from the stock, differs from laying, inarching, and grafting, inasmuch as the ascending or wood sap is in no way concerned. The bud is fixed entirely by the sap of the bark, with probably some aid from what may ooze out of the soft wood.

But the scallope or French mode of budding, in which a section or scallope of bark and wood containing a bud is taken from one tree, and applied to a part of the stem of another tree, where a similar scallope had been removed, acts exactly on the principle of the graft, and is best performed in the grafting season.

The common mode of budding, which is the simplest, and with most plants the most certain, is performed when the young shoots of the season have nearly completed their growth; then the buds are formed, and the bark separates freely from the young wood. Scallope budding is chiefly used in propagating roses, as in them the bark often does not separate readily; but in greenhouse plants it might be used in the case of plants which are too small for separating a single bud, but where a scallope containing two or three buds might be taken off and inserted. In this way more plants might be got from a given extent of shoot, than either by grafting, inarching, or laying, or even propagating by cuttings.

Formerly it was thought grafting and budding could only be applied to ligneous plants; but now it is found that annuals, biennials, and perennials, may be budded or grafted. The Dahlia is sometimes propagated in this way : a young shoot of any rare sort being grafted on a tuber of a hardy kind. Baron Tschoudi, an ingenious French horticulturist, informs us that he has grafted the love-apple on the potatoe, the melon on the gourd, and the cabbage on the cauliflower. (See *Essai sur la Greffe de l'Herbe, par le Baron Tschoudi, &c.* 1819.)

Subsect. 5. *Propagation of Green-house Plants by their leaves.*

The propagation of plants by their leaves is comparatively a recent discovery. It takes place in nature with a few plants, such as the *Dionæa*, Sea-onion, &c., and with the hairy Ladies'-smock (*Cardamine hirsuta*), a native of this country. All the improvement which art has hitherto made on this natural mode of propagation, is to take off the leaves when they are fully grown, and instead of dropping them on the earth at random, to lay them on their backs on moist earth shaded either by position or by covering, or, what is generally preferable, by a little fresh moss strewed over them. All the culture necessary till the young plants form themselves on the edge of the leaves, is to keep the moss and earth below of a regular

degree of moisture, and exclude bright sunshine.
The following green-house and hot-house plants
may be propagated in this way, covering the more
delicate with a bell-glass:

Bryophyllum calycinum.

Gloxinia speciosa.

Dionæa muscipula.

Verea crenata.

Arum, various species.

Hoya carnosa.

Aloe, most of the species.

Xylophylla, any species.

Crassula, many species.

Cotyledon orbicularis.

All the tunicate and squamose
bulbs, and probably many
others.

Other plants throw out roots from the petioles
of the leaves, without, as far as has hitherto been
observed, having the power to generate buds.
Such are *Camellia, Aucuba, Laurus,* &c. Others
root from the points of their lateral branches,
without, as far as has been observed, having the
power to form a main shoot, as the *Auracaria* or
Brazilian Pine.

We have already (page 116) described the
mode in which scaly bulbs may be increased by
the leaves, viz. by pulling a full-grown leaf off
with the scale attached, and planting it in a pot of
sandy loam, when after a short time the leaf
throws down the sap which should have gone to
the parent bulb, and this forms a callosity of gra-
nulous matter which soon takes the form of a
young bulb or bulbs.

237

Tunicate bulbs we have also shown may be increased by cutting off the upper part of the bulb horizontally, when each coating will throw out little gems from its upper edge (page 116).

These and other modes of propagating plants are partly the result of accident, and partly of the great progress made in vegetable physiology. The more common modes of *dividing the roots* of tuberous or fusiform plants, and *dividing at the root* of fibrous perennials, or under-shrubs, are so simple and well known as not to require description.

SECT. VII. *Of the Management of the Conservatory.*

The treatment of plants in a conservatory differs in nothing essential from that of plants in a greenhouse : the former being fixed in the soil, in order to give them the advantage of the natural weather, the superstructure must be moved from them, or at least its sashes taken away; while the latter being portable may be moved from the house.

All the directions as to routine culture for the different seasons, are alike applicable to the conservatory. The same temperature is required, the same light, the same circulation of air, and the same liberal supply of water to plants in a growing state, and scanty watering to those which

are at rest or dormant. Weeds and insects are
to be removed, and all dead, decayed, withered
or damped leaves picked off as soon as they appear. The plants must be neatly pruned, thinned,
and trained to handsome shapes by the aid of
sticks and threads, and, where the branches are
very strong, by copper wires attached to fixed
points in the structure, or strong sticks inserted in
the soil. Climbers are generally more numerous
in a conservatory than in a green-house, and they
must be particularly attended to so as to make
them hang in graceful festoons, and yet so thinned
and pruned as to keep them in a flowering state.
It is a common idea that only fruit trees require
pruning; but the fact is, that trees or shrubs cultivated for their blossoms require just as much pruning as when the blossoms are to be succeeded by
edible fruits. Conservatory trees therefore must
be pruned just as regularly every autumn or spring,
as are fruit trees. The best season for pruning
is when the tree is not in a growing state, nor the
sap much in motion; the next best is when the
tree is in full growth, and has produced its shoots
nearly their whole length; and the worst is when
the sap is descending after growth is completed,
or ascending when it is commencing.

The end of May or beginning of June is the
usual time of exposing the plants of the conser-

vatory fully to the weather. Previously to that, the front sashes should have been removed for some weeks, and air admitted by the roof sashes night and day for at least a week. This hardens the plants by degrees, and prepares them for a full exposure to the weather.

About the middle of September, or as soon as it is thought necessary to house the other green-house plants, the roof lights of the conservatory should be put in their places; the ends may remain open a week or ten days, and the front a fortnight or three weeks longer according to the weather. Afterwards, when the house is shut up every night, the greatest care must be had to ventilate freely every day; for the damps of this foggy season are very apt to generate mouldiness on the soil, and on the leaves and shoots not freely exposed to the light and air.

A conservatory after being planted a few years will require the liberal use of the knife to keep the more woody plants within due bounds; most of the shrubs will require to be cut in and pruned so as to keep them clothed to the surface, and the trees will require to be headed down to the tops of their stems. Both trees and shrubs will occasionally require to be cut down to the ground, and when this is done care must be taken not to go below the bark. We have heard of a collection of Camellias planted in a conservatory in Kent, which after seven years growth were cut down in this way to renew them : they all grew most vigorously, but when they came to blossom showed only single red flowers, which told what had been done. The conservatory alluded to had a fixed roof, and the plants could only enjoy the weather through the sides ; by which most erroneous construction the plants became etiolated, and so naked below that every six or seven years they had all either to be cut down or renewed.

A very agreeable ornament to the beds and borders of the conservatory is formed by planting them with the stronger-growing species of *Ixiæ*, *Gladioli*, and the other genera lately separated from these. The fine tall stems of these bulbs shoot up through the woody undergrowths, and their brilliant red, scarlet and white flowers make a fine appearance. They may be planted in the beds exactly at the same time in which they are potted for green-house culture, and they may remain two or three years in the soil, if desired, without injury, flowering every year.

A cistern is frequently placed in the conservatory and devoted to the culture of aquatics. Though most of these are hot-house plants, and require more heat than is proper for this department, yet some some species of the *Nymphæa*,

Menyanthes, Aponogeton, Hedychium, Calla, Trapa, &c., will live there, and flower with a moderate degree of vigour. These plants require to be potted in rich loamy soil in moderate-sized pots; which in the case of plants with long upright foot-stalks to the leaves should be placed a foot or more under water; while others with slender floating footstalks, as *Menyanthes*, should be set on shelves or other supports so as to raise the brim of the pots to within an inch or two of the surface. A small but daily supply of fresh water, and occasional clearing from slime or other dirt that happens to accumulate, is all the culture such plants require. They are generally increased by dividing at the root; and some few, as the *Trapa*, ripen seeds.

SECT. VIII. *Management of Town Green-houses, and of Plants in Chambers at Routs.*

Town green-houses are generally on a small scale, and seldom accompanied by sheds for potting and shifting, stocks of mould, and a platform for setting out the plants in summer: if they were, it is probable these adjuncts would be of very little use, for no green-house plant will ever thrive in a town where fossil coal is generally consumed. All that can be said with advantage on town

R

green-houses might be comprised in very few words: viz. that the only way to have them look well is to agree with a nurseryman to keep up a supply of verdant flowering plants for such a part of the year as the family is in town. We are confident there is no other mode that will be attended with success, till the nature of plants or the nature of a coal fire is considerably altered.

A number of London green-houses placed behind the houses on the tops of kitchens and other offices, and of plant cabinets communicating with living-rooms, are maintained in order by nurserymen in this way; and a number also are kept in order, as it is called, by jobbing gardeners, who call occasionally to see that the plants are properly watered, who supply pots of mignionette, and who shift the plants in spring, and prune them in autumn. But the green-houses managed in the latter mode are wretched vegetable abodes,—hospitals or pest-houses of plants; and to any person who knows what a green healthy plant is, they are deformities rather than ornaments.

Another mode in which green-houses in the metropolis are sometimes managed, is as follows: The occupier of the London house has a villa within 15 or 20 miles of town, where he has a green-house, or grapery and green-house combined,

with pits and hot-beds, and keeps a gardener.
From the country weekly supplies of vegetables,
butter, eggs, fowls, cream, and pots of forced articles, are obtained, and faded flowers and sickly
green-house plants returned. This is certainly the
proudest and most gratifying mode of all; but yet
as far as the green-house is concerned, it is attended
with less show than where a nurseryman of extensive practice in the culture of plants in pots, is
employed. Such a nurseryman has four grand
sources of disposing of plants in flower, besides
his ordinary chance buyers and private custom:
first, he sells to the hawkers; next, he can send to
Covent-garden market by cart-loads; third, he
supplies green-houses by the month or year, and
lastly, he supplies routs. He is never therefore
without a large stock in hand; and in order to
make the most of these, he first tries them at market when they are barely coming into flower,
which state suits the shopkeepers who buy to
keep them a week or two to sell again. If he
fails there, the hawkers come to him every morning and see what he has got a bargain; and his
next resource is the green-houses, cabinets, or
chamber-stages, which he supplies by contract;
and when he removes them from thence in their
last stage beauty, he sends them to a rout, where
one night in general kills or half kills alike the

best and worst of plants; and for which he gets
the plant returned and half its price. In this
way it is that a public dealer can always afford
to keep up a finer display of plants in a town
green-house than any private gentleman whatever
with a country villa.

Some families who have no green-house, keep
up a stock of plants in their rooms by purchasing
at the nurseries, and at Covent-garden market;
but this is done at greater expense than by contracting with a nurseryman, because the family
can make no use of the plants when out.of flower.
When this mode is confined to annuals, such as
mignionette, wall-flowers, sweet-peas, &c., or to
bulbous roots, it does very well, as these when
they have done flowering are of no use to any one.

But notwithstanding all that we have said, there
is no doubt a number of persons who will go on
managing their town green-houses themselves,
and to such it is proper we should offer the best
advice in our power.

1st. We recommend them to grow only the most
hardy plants, and such as flower freely, as Geraniums, Myrtles, Coluteas, Pittosporums, Corræas,
Acacias, and the like; and to indulge only in the
hardier heaths and camellias, with but a few
orange trees.

2d. We advise pots more than usually well

drained; say none but what have their bottoms covered with an inch of gravel at least, to prevent all chance of the stagnation of water, and lessen the evils of what is almost certain of taking place, over-watering.

3d. Whatever soils may be used, we would advise mixing some very coarse sand and small gravel with them to prevent the soil, when watering was too long deferred, (which it would certainly be every now and then,) from getting so hard and compact that water would not penetrate through it.

4th. Where saucers are placed below the pots to prevent the water which escapes from the latter from wetting the floor of the room or green-house, we advise the saucer to be half filled with gravel, in order that the soil or gravel in the inside of the pot may never stand soking in water. When this is not done, then as a substitute let a large sponge be used an hour after every watering, to suck up the water which has run through into the saucers : but few servants can be trusted to attend to this.

5th. We advise frequent watering over-head to remove dust and dirt; and when this is not done effectually by a shower from the rose of the water-ing-pot held high above them, or from the free use of Read's syringe, then we recommend wash-ing each plant leaf by leaf with a wet sponge.

6th. We advise giving as much air as possible

in the day time, and on account of the frequent waterings over-head that will become necessary, and also to keep the plants in a fresh, verdant, and even growing and flowering state during winter, a rather higher degree of heat at that season than we have recommended for the villa green-house.

7th. The construction of the green-house should be such as to admit as much light as possible—more so if practicable than the villa green-house, as the light of cities is less pure. The plants should be set nearer to the glass for the same reason; and, what is seldom necessary in the country at all, the glass should be cleaned three or four times a year. In correspondence with these directions, few creepers are admissible in town green-houses.

8th. The plants should, if room and other circumstances permit, be completely exposed to the open air in the same way as green-house plants, and shifted, and in all other respects not above noticed, treated like them.

The management of plants in chambers can only be understood to apply to the short time in which they are kept there. This time should be as short as possible, if it be intended that the plants should live and thrive afterwards. The only true way to have a fine display on the cham-

247

ber-stage, is, never to bring the plants there till
they are just coming into flower, and to remove
them when the flowers first show indications of
decay, unless the plant appears to be growing
sickly before, which with heaths, geraniums,
and camellias, is very often the case. During the
time they are kept in the chamber, the surface of
the pot should either be covered with fine fresh
moss or coarse sand, to lessen the evaporation
from the moist earth of the pot, and to prevent
the earth from caking with the heat of the chamber : the water which passes through the bottom
of the pot into the saucer, should be instantly
extracted with a sponge to prevent evaporation,
and none should be spilt on the leaves or the
stage. As much air should be admitted by opening the windows, as is consistent with the use of
the room, and the stage should, in the day time
at least, be kept as near the windows as possible.
It is not essential that the sun should shine on the
plants, so as they have abundance of reflected
light. It is almost needless to observe, that they
should before being brought there be tied up, or
otherwise arranged with the greatest neatness,
and that while in the chamber, all decayed or
injured parts should be removed, and any dirt or
dust carefully wiped off with a moistened sponge.
When the plants are taken back to the green-

house, a little extra heat and moisture will in general recover them.

Plants at routs require little management while there, but must be tastefully arranged individually by rods and threads, and well syringed, and also watered at the root. The soil should be covered with moss, and the pot either cleaned and painted in any appropriate body colour, or chalked, or covered with coloured paper. An earthen brown with black and gray lines is among the most suitable colours, whether for the temporary painting, chalking, or papering. Where the plants are only to remain one night, they need not be set in saucers, but only on paper or small carpets of the size of the bottom of the pot; but where they are to decorate the apartments for two or three days in succession, they should be set in saucers on a little gravel, and over the gravel the saucer filled brimful with moss or fine green turf. This looks well, and the space occupied by the gravel admits of giving the plants daily a little water, which greatly refreshes them in this state of trial.

The arrangement of the plants in the rooms is various, and depends on the kind of rout or entertainment. In common cases they are placed in recesses, and on side-tables, and near glasses which may reflect them; and a few choice specimens are scattered over the floor as single objects. But in more select entertainments, a proportionate attention is paid to their arrangement. During dinner a few pots of fruit-bearing shrubs, or trees with their fruit ripe, are ranged along the centre of the table, from which, during the dessert, the fruit is gathered by the company. Sometimes a row of orange trees, or standard peach trees, or cherries, or all of them, in fruit, surround the table of the guests; one plant being placed exactly behind each chair, leaving room for the servants to approach between. Sometimes only one tall handsome tree is placed behind the master, and another behind the mistress; and sometimes only a few pots of lesser articles are placed on the side-board, or here and there round the room.

The drawing-room is sometimes laid out like an orange-grove by distributing tall orange trees all over it in regular quincunx, so that the heads of the trees may be higher than those of the company : seats are also neatly made over the pots and boxes, to conceal them, and serve the purpose of chairs. One or two cages with nightingales and canary-birds are distributed among the branches, and where there is a want of real fruit, that is supplied by art. Sometimes also art supplies the entire tree, which during artificial illumination is hardly

recognised as a work of art, and a very few real
trees and flowers interspersed with these made
ones, will keep up the odour and the illusion to
nature.

Sometimes large picture galleries are laid out
in imitation of parks in the ancient or modern
style, with avenues or with groups and scattered
trees. At masqued routs, caves and grottos are
formed under conical stages, and covered with
moss and pots of trees, in imitation of wooded
hills. In short, there is no end to the arrangement
of plants at routs: and the reader is not to suppose that only real plants with roots are necessary for this purpose; for, provided a few of these
be judiciously introduced, all the rest can be
effected by branches of box, laurustinus, laurel,
juniper, holly, &c., decorated with artificial flowers and fruits, and fitted to stems or trunks to
answer either as trees or shrubs; and besides
these, whole pine and fir trees, the spruce especially, can be cut over, and thus admirable groves
formed in a short time. Artificial supplies of
odour of the rose, the orange, or the jasmine, are
readily supplied. Much romantic splendour may
be produced in this way with little expense of
green-house plants.

Next to the common domestic flower-stage, and
a handsome plant placed here and there in spare

places in the lobby, hall, staircase, &c., are a few
choice specimens of tall plants in fruit or flower,
distributed in the drawing-room : the orange, the
camellia, the acacia, and tree heaths, are well
adapted for this purpose.

Sect. IX. *Management of Bulbs in Water-
glasses ; care of Nosegays, &c.*

The next point for consideration is the management of bulbous-rooted flowers planted on water.
This is done in various ways. Sometimes a large
vessel 2 or 3 feet in diameter, and a foot deep,
has a cover fitted to it; in this cover are holes
(the largest towards the centre) in concentric
circles, on which to place a collection of bulbs,
from the crocus, which is placed outside, to the
strongest polyanthus, narciss, which when in
flower forms the apex of the cone in the centre.
Sometimes a cone, or semi-globe, or semi-dome, is
formed by tin troughs 6 or 8 inches deep, and not
wider than 2 inches, to which covers with holes
for the bulbs are fitted. This construction admits
of placing bulbs in horizontal rows rising above
one another to the apex of the cone or dome.
The bulbs may either be mixed, or of the same
species but of different varieties, as a cone of

many different sorts of hyacinths, another of narcissi, tulips, &c. The cases being filled with
water, and the bulbs placed over the holes, the
whole cone should then be neatly covered with
green moss, so as to show only a part of the necks
of the bulbs. They will grow beautifully; and
when the flower-stalks can no longer support
themselves, there is a wire ring to each trough
about 6 inches above its surface, and another 3
inches above that, to which they as well as any
weak leaves can be tied. The water in each
trough can be renewed, without disturbing the
bulbs, by a small cock which points to the inside
of the cone or dome; and to the whole there is a
tin bottom to collect any drops of water which
may escape from leakage or otherwise. The in-
vention of this mode is by some attributed to the
Dutch, and by others to a Russian; it having
been chiefly displayed in some of the Petersburg
palaces.

The most general mode of planting bulbs on
water is by flower-glasses, well known to every
one. There are two sorts; one with darkened
glass, which is the best for the roots as excluding
the light; and the other with bright glass, which
shows to the spectator the progress of the roots.
Here each bulb is grown by itself, and when its
flower, stem, or leaves require support, one or

more props are affixed to the base in which the glass is placed.

All sorts of bulbs may most probably be blown on water; but those chiefly so blown are the hyacinth, polyanthus, narciss, early tulip, Persian iris, crocus, and a few others. The colchicum, Guernsey lily, saffron crocus, and some other autumn bulbs, will also flower very well in this way : the variegated colchicum, when it is prolific in flowers, has a very fine appearance.

The season for placing the bulbs on water may be any period after they have been matured ; but the most usual time with spring bulbs is October, and from that month to February; and with autumn bulbs August and September. We have already mentioned that planting in earth for a few weeks such bulbs as are to be blown on water, is the best mode of causing them to protrude roots freely, which when they are placed on water at once is not always the case. Whenever the roots are a quarter of an inch in length, take them out of the earth, wash them gently so as not to injure the radicles, and then place them on the water.

It is not essential that bulbs on water should be placed in much heat, for the principal stimulus to a new planted bulb is the moisture; and if the room in which the glasses are placed be kept to 45° or 48°, that will promote their vegetation for some time as much as 10° or 15° higher. When the flower-stem has risen an inch or two, then the heat may be considerably increased: that is, the glasses may be moved from a room without a fire to one where a fire is kept, and where the temperature will generally be found between 55° and 65°. Here they will advance with considerable rapidity, especially if placed on a stand or stage near a window of south or south-east aspect. They will blow however without any sun ; but the colours of the flowers will be inferior. It is a remarkable circumstance of the crocus, that it keeps its petals expanded during a tolerably bright candle or lamp light, in the same way as it does during the light of the sun. If the candle be removed, the crocuses close their petals, as they do in the garden when a cloud obscures the sun; and when the artificial light is restored, they open again as they do with the return of the direct solar rays.

Those who keep bulbs on water are often at a loss when to change it. There is no fixed time for this purpose: the principle is to keep the water sweet and pure. In a temperature of 45° or 48°, when the bulbs are newly planted, this will be effected by changing once a week: at 60°, and the glass nearly filled with roots, the water will get putrid and show a muddiness in two or three days or less, and whenever it does so it ought to be changed. The operation of changing is easily done by one person, when the roots are only an inch or two long ; but after the flower-stems are of some length, and the roots nearly at the bottom of the glass, two persons become requisite ; one to take out the bulb and hold it, and to dip its roots once or twice in a vessel of clear water to clean them a little ; and another to empty and rinse out the glass and refill it with water.

It is essential that the water used for renewal or for rinsing the roots, should be of the same temperature as that which it is to replace ; and this can easily be done by keeping it a day or two in a room of similar temperature, or pouring a little hot water into cold water, and proving it with a thermometer. Whether the water be hard or soft is of no great consequence, but soft or rain water is considered preferable.

Forced bulbs are seldom good for any thing afterwards. However, those who wish to preserve them, may immerse them wholly in water for a few weeks, and then having taken them up and dried them in the shade for a few days, they may be planted in good soil, when they will sometimes flower the second year. It does not clearly appear in what way the water operates when the bulb is wholly immersed ; but it is certain that

bulbs so treated increase in size and solidity by
it, and have an incomparably better chance of
flowering the second year, than those which have
not been so treated.

Most probably their total immersion enables
them to obtain a greater proportion of oxygen
from the water.

Nosegays should have the water in which their
ends are inserted changed on the same principle
as bulbous roots ; and a much faded nosegay, or
one dried up, may often be recovered for a time,
by covering with a crystal bell or cap, or by sub-
stituting warm water for cold.

Those who wish further details as to plants in
rooms, or what the French and Germans call window gardening, may consult a little work entitled
Le Jardinier des fenêbres, des apartemens, &c.
Paris, 1823, 12mo.

END OF PART I.

PART II.

THE
GREEN-HOUSE CATALOGUE;

CONTAINING

A BOTANICAL ARRANGEMENT OF ALL THE GREEN-HOUSE
PLANTS IN CULTIVATION, WITH THEIR PROPER SOILS,
MODES OF PROPAGATION, REFERENCES TO BOTANICAL
WORKS IN WHICH THEY ARE FIGURED; AND OTHER
PARTICULARS.

PART II.

——————

THE

GREEN-HOUSE CATALOGUE,

INCLUDING ALL THE GREEN-HOUSE AND FRAME PLANTS
HITHERTO IN CULTIVATION.

————

IN arranging this Catalogue we have adopted
the natural method of Jussieu, for two reasons
that will at once appear obvious to the botanical
cultivator, though they may require some expla-
nations to the general reader.

In the artificial system of Linnæus, plants are
brought together in orders according to the number or position of the stamens and pistils of the
flower, without regard to any thing else; and
as plants which are alike as to stamens and pistils are often exceedingly unlike in every thing
else, there is no sort of harmony or resemblance in the general appearance of any Linnæan

(B) 2

order. But in the system of Jussieu plants are brought together into orders and groups, not from their agreeing in stamens or in any other particular part of the plant, but from their agreeing in the greatest number of particulars; and hence a general harmony and resemblance pervades the whole order, and any person who knows or can recollect the appearance of any one plant in that order, will have a tolerably correct idea of the whole group. The names of these orders are generally taken from some genus contained in it, which is reckoned a prototype of the whole; and thus to those who recollect any species of that genus a very useful practical idea of the order will be formed. Thus, whoever knows any species of the genus *Myrtus*, Myrtle, will be able to form a pretty good idea of the order *Myrteaceæ* and so on. This is our first reason for arranging our catalogue according to the natural method, rather than according to that of Linnæus, or of any other mode whatever.

A second reason is, that plants of the same natural order very often agree as to their modes of artificial propgaation and culture: many even agree in their natural modes of propagation; and with some orders, as the *Amaryllideæ, Rhodoraceæ, Geraniaceæ, &c.*, it is thought that mules may be formed between any two species belonging to

any one of these orders, though of different genera. Hence, by treating of the genera of these
orders together, much repetition is avoided, and
the reader enabled to have a much clearer idea of
what he is reading, by always bearing in mind
the leading features of the plants of the order.
Thus, the order *Irideæ*, which contains a considerable number of genera, almost all bulbs, growing in the same soil, and propagated by offsets, if
arranged alphabetically, would afford a genus for
most letters of the alphabet, and thus be scattered
throughout the whole catalogue. Then to each
genus the soil, mode of propagation, and the
statement that it was a bulb, &c. &c., must have
to be added; whereas, by keeping them together,
the title of the order at once gives the idea of the
plants contained in it: their ensate leaves, brilliant
flowers, and bulbous roots. These ideas the reader
will carry along with him in perusing the names
and descriptive traits of each genus and species,
and thus have a more definite notion of what he
is reading about than could be otherwise obtained.

An arrangement according to the natural re-
semblances of plants is also far more suitable for
such as wish to choose a general collection; for
to a person who does not know plants, what guide
will the mere names afford? or their alphabetical
or Linnæan arrangement? None whatever! But

a person wholly unacquainted with plants, if he chooses a species or one or two species from each of the natural orders, even at random, would be certain of having a collection exhibiting a prototype of the whole list of green-house plants. If instead of this he were to select two or three plants from each of the Linnæan orders, he might omit many of the natural orders altogether; might omit some of the finest kinds of vegetable beauty, and of course could not have any thing like a complete collection. In short, the advantages of the natural arrangement of plants are more numerous than we can here afford room to explain; not only to such as already know plants, but even to those who are ignorant of botany. Some bigoted and interested admirers of the Linnæan system have long tried hard to prevent the spread of that of Jussieu in this country; but it has finally prevailed; and after the long-continued exertions in its favour by Mr. Brown and Mr. R. A. Salisbury, we at last observe that even Sir J. E. Smith, the "possessor of the Linnæan herbarium," has annexed the names of the natural orders of Jussieu and Brown to his translation of *Flora Britannica.*

We are far from insinuating by these observations that the natural method will supersede that of Linnæus; it is neither desirable nor probable that it ever should do so. The method of

Linnæus is by far the best for a beginner; it facilitates the knowledge of plants as individual objects, while the natural mode enlarges the understanding by generalizing facts. " Plants arranged according to the natural method," it has been observed, " may be compared to words arranged according to their roots or derivations; arranged according to the Linnæan method, they may be compared to words in a dictionary."

The alphabetical mode of arrangement is that which is commonly resorted to in catalogues or lists of this kind: but this is the worst of all modes, since the mere circumstance of agreeing in the initial letter of the name, can never be any philosophical or scientific ground of union; and as to the convenience of turning to any genus when they are so arranged, that is much more completely and effectually obtained by a general alphabetical index to the whole work, which under the name of any one genus refers not only to the catalogue but to all the different parts of the book where that genus is mentioned.

Such are the reasons for the arrangement we have adopted; which being rather new in works of culture, we deemed it necessary thus to explain its uses and advantages.

RANUNCULACEÆ.

PÆONIA *Moutan*, the Chinese Tree Pæony, a low shrub, a native of China, introduced in 1789, and flowering in April, May and June.

There are three varieties, viz.

P. *M. papaveracea*, the poppy-flowered, A. R. C. ic. B. C. 547.
    banksia,   the common,   B. M. 1154.
    rosea,   the rose-coloured.

These are frame plants, and will even bear our ordinary winters in the open air if sheltered by a wall or by a wicker cap: but the true way to have them flower in perfection is to plant them in the front of a conservatory, where they will make a fine show in spring. They also make a fine appearance trained on a south wall or any warm part of a house along with the *Rosa semperflorens*. They grow in any rich light soil, and are propagated by ripened cuttings, in the shade, and without cover.

ATRAGENE *capensis*, Cape Atragene, B. M. 716, a shrub of no great beauty, introduced from the Cape of Good Hope in 1795, and flowering in March and April. It grows in loam and peat, and is generally increased by seeds.

CLEMATIS *calycina*, Minorca Virgin's-bower, B. M. 959, a climbing shrub introduced in 1783 from Minorca, and flowering in February and March.

C. *aristata*, awned-anthered Virgin's-bower, B. R. 238, B. C. 620, a climbing shrub introduced from New Holland in 1812, and flowering in May, June and July.

C. *brachiata*, armed Virgin's-bower, B. R. 97, a climbing shrub from the Cape of Good Hope, which flowers from October to the end of the year.

C. *Massoni*, Masson's Virgin's-bower, a climbing shrub introduced by Mr. Masson from the Cape.

These climbers are all rapid growers, and will soon overrun a green-house if not kept under by the knife, or, what is better, confining their roots to pots. They grow in sandy loam or indeed in any soil, and propagate freely by cuttings under a glass cover.

KNOWLTONIA *rigida*, thick-leaved Knowltonia, a perennial herbaceous plant, introduced from the Cape of Good Hope in 1780, and flowering in March and May.

K. *vesicatoria*, blistering Knowltonia, B. M. 775, a perennial introduced from the Cape in 1691, and which flowers in February and April.

These are plants of no beauty, but free growers in loam and leaf-mould, and they are propagated like other herbaceous plants by dividing the root. They also ripen seeds.

MAGNOLIACEÆ.

HIBBERTIA *volubilis*, twining Hibbertia, A. R. 126, B. M. 449, a twining shrub introduced from New South Wales in 1790, and flowering from May to October.

H. *grossulariæfolia*, gooseberry-leaved Hibbertia, B. M. 1218, a New Holland shrub introduced in 1803, and flowering from March to August.

These plants grow vigorously in loam and peat or leaf-mould with a little coarse sand, and cuttings root in sandy loam under a hand-glass.

ILLICIUM *floridanum*, red-flowered Aniseed tree, B.C. 209, B. M. 439, a frame shrub, introduced from Florida in 1766, and flowering in April, May, and June.

I. *parviflorum*, yellow-flowered Aniseed tree, a frame shrub, introduced from Florida in 1790, and flowering in May and June.

These are handsome evergreens of easy culture in light loamy soil, and with the protection of glass during winter covered with mats. They do better and flower freely, however, in the green-house or conservatory. They are commonly increased by layers, but will also grow by cuttings, planted in sand under a glass.

MAGNOLIA *conspicua*, Youlan Magnolia, B. M. 1621, a frame tree, introduced from China in 1789, and flowering from February to April.

M. *obovata*, purple Magnolia, B. M. 390, a frame tree, introduced from China in 1790, and flowering from May to July.

M. *tomentosa*, slender Magnolia, a frame tree, introduced from China in 1804, and flowering from March to May.

M. *pumila*, dwarf Magnolia, B. M. 977, a green-house tree, introduced from China in 1786, and flowering all the year.

M. *fuscata*, brown-stalked Magnolia, B. M. 1008, a green-house tree, introduced from China in 1789, and flowering in April and May.

M. *annonæfolia*, small-flowered Magnolia, a green-house tree, introduced from China in 1804, and flowering in April and May.

This is a noble genus of plants; and though the most magnificent species, as M. *grandiflora, glauca, auriculata*, &c., are hardy trees, yet those kept in the frame and green-house have fine large white flowers. The soil best suited for the Magnolia is loam and peat with a little sand. The species are generally propagated by laying or by inarching the more rare on the more abundant kinds. Thus the green-house and frame sorts are often grafted on M. *purpurea* and *obovata*. Some of the slender-

wooded species, as M. *pumila* and *fuscata*, will root from
ripe cuttings in sand under a glass.

MENISPERMEÆ.

SCHISANDRA *coccinea*, scarlet-flowered Schisandra,
B. M. 1413, a green-house shrub, a climber, introduced
from North America in 1806, and flowering in June and
July.

This is a showy plant, which grows freely in sandy
loam, with a little peat or leaf-mould. It may be increased by layers or ripened cuttings, planted in sand
under a glass.

CISSAMPELOS *capensis*, Cape Cissampelos, a green-
house tree, introduced from the Cape of Good Hope in
1775. It is a plant of little beauty, but of easy culture,
and propagated by cuttings in loamy soil.

CRUCIFEREÆ.

IBERIS *semperflorens*, broad-leaved Candy-tuft, an evergreen undershrub, introduced from Sicily in 1679, and
flowering in white umbels all the year.

I. *gibraltarica*, Gibraltar Candy-tuft, B. M. 124, an
evergreen undershrub, introduced from Spain in 1732,
and flowering in May and June.

I. *ciliata*, ciliate-leaved Candy-tuft, B. M. 1030, an
evergreen frame perennial, introduced from Caucasus in
1802, and flowering in June and July.

These are ornamental little plants, and valuable as
evergreens: they thrive in any light soil or brick rubbish, and are readily increased by cuttings in the same
soil, and under a hand-glass.

CRAMBE *fruticosa*, Madeira Colewort, a low spongy

undershrub, introduced from Madeira in 1777, and flowering from May to November.

CRAMBE *strigosa*, Canary Colewort, a low spongy undershrub, introduced from the Canaries in 1779, and flowering in May and June.

These are plants of no beauty. They grow in sandy soil, and may be raised from seeds or by cuttings under a glass in the same soil.

BISCUTELLA *sempervirens*, downy-leaved Buckler Mustard, a low frame shrub, introduced from Spain in 1784, of no beauty whatever, but easily increased by seeds or cuttings in any sandy soil; it flowers in May and June.

VELLA *Pseudo-Cytisus*, shrubby Cress Rocket, a low spongy frame shrub, introduced from Spain in 1759, and flowering in April and May. It grows in any dry rubbish, or limy soil, and is readily increased by cuttings or seeds.

LEPIDIUM *divaricatum*, close-spiked Pepperwort, a low shrub, introduced from the Cape of Good Hope in 1774, and flowering from May to August.

L. *Cardamines*, Spanish Cress, a frame biennial, introduced from Spain in 1789, and flowering in June and July.

L. *subulatum*, awl-leaved Spanish Cress, a spongy undershrub, introduced from Spain in 1739, and flowering in July and August.

These are plants of no beauty whatever; they grow in any light soil, and are increased by cuttings in the same soil or by seeds.

ALYSSUM *spinosum*, thorny Mudwort, a very low evergreen frame shrub, introduced from the south of Europe in 1683, and producing showy yellow flowers in July and August.

FARSETIA *Cheiranthoides,* Stock Farsetia, an undershrub, with spongy wood not unlike the Wallflower, introduced from the Levant in 1788, and flowering in June
and July. It grows in light soil, seeds freely, or may be
increased by cuttings like *Cheiranthus Cheiri.*

HELIOPHILA *incana,* hoary Heliophila, a low shrub,
introduced from the Cape of Good Hope in 1774, and
flowering from May to July.

H. *filiformis,* awl-podded Heliophila, an annual, introduced from the Cape of Good Hope in 1786, and
flowering in July and August.

H. *platysiliqua,* broad-podded Heliophila, an undershrub, with spongy wood, introduced from the Cape of
Good Hope in 1774, and flowering in July and August.

H. *coronopifolia,* Buck's-horn-leaved Heliophila, a
green-house biennial, introduced from the Cape of Good
Hope in 1778, and flowering in June and July.

These, unlike most of the plants of this order, have
some pretensions to beauty: they are easily preserved in
light soil, and increased by seeds or cuttings.

CARDAMINE *africana,* African Lady's Smock, a perennial, introduced from the Cape of Good Hope in 1691,
and flowering in May and June. It grows in light
soil, as loam and leaf-mould, and is increased by dividing
at the root, or by slips or cuttings in the same soil.

SISYMBRIUM *millefolium,* Milfoil-leaved Sisymbrium,
a low spongy-wooded green-house shrub, of no beauty,
introduced from the Canaries in 1779, and flowering
from May to September. It grows in sandy loam, and
is easily increased by seeds or cuttings.

NOTOCERAS *canariensis,* an annual, introduced from
the Canaries in 1779, and flowering in August and September. It grows in light soil, and is increased by cuttings.

CHEIRANTHUS *strictus*, upright Wallflower, a spongy-wooded undershrub, introduced from the Cape of Good Hope in 1802, and flowering in July.

C. *mutabilis*, changeable Wallflower, B. M. 195, also a low shrub, with soft herbaceous-like branches, which was introduced from Madeira in 1777, and flowers from March to May. It is esteemed in collections for its early flowers and hardy nature as to soil.

C. *longifolius*, long-leaved Wallflower, a low herbaceous-like shrub, introduced from Madeira in 1815, and flowering from September to January.

C. *frutescens*, entire-leaved Wallflower, a low shrub, introduced from Teneriffe in 1815, and flowering in March, and till June or July.

C. *scoparius*, crowded-branched Wallflower, B. R. 219, a low shrub, introduced from Teneriffe in 1815, and flowering from March to August.

C. *linifolius*, flax-leaved Wallflower, a low green-house shrub, introduced from Spain in 1815, and flowering from March to August.

C. *semperflorens*, ever-blowing Wallflower, a low shrub, introduced from Barbary in 1815, and flowering all the year. It is a very desirable accession to a green-house where neither heaths nor camellias are introduced.

C. *silenifolius*, Catchfly-leaved Wallflower, a low green-house shrub, by some considered only a variety of the last, which flowers all the year.

C. *tenuifolius*, fine-leaved Wallflower, a low shrub, introduced from Madeira in 1777, and flowering in May and June.

All these plants grow readily in any light soil, and are as easily increased by cuttings as the common wallflower; to which none of them can be considered as any thing like equal in beauty or fragrance.

MATHIOLA *fenestralis*, Window Stock, a biennial plant, a free flowerer in July and August, and known since 1759.

M. *odoratissima*, Persian Stock, B. M. 1711, a low spongy shrub, a native of Persia, introduced in 1797, and flowering in May, June and July.

M. *tristis*, dark-flowered Stock, B. M. 729, a low spongy shrub, introduced from the South of Europe in 1768, and flowering from May to July.

These plants are of easy culture in any light soil, and increase readily by seeds or cuttings.

SINAPIS *frutescens*, shrubby Mustard, a low spongy-wooded shrub, a native of Madeira, introduced in 1777, and flowering from December to June. It is a plant of no beauty, but of the easiest culture in light soil, and propagated by seeds or cuttings.

CAPPARIDEÆ.

RESEDA *glauca*, glaucous Reseda, a perennial, introduced from the South of Europe in 1700, and flowering in May and July.

R. *dipetala*, flax-leaved Reseda, a biennial, a native of the Cape of Good Hope, introduced in 1774, and flowering in August.

R. *scoparia*, broom-like Reseda, a low spongy shrub, introduced from Teneriffe in 1815, and flowering in the green-house in August and September.

R. *fruticulosa*, shrubby Reseda, a low herbaceous-like shrub, a native of Spain, introduced in 1794, and flowering in September.

R. *bipinnata*, bipinnate-leaved Reseda, a low herbaceous-looking shrub, introduced from Spain in 1816, and flowering from June to August.

R. *odorata*, var. *frutescens*, the tree Mignonette,

B. R. 227, an odoriferous ornamental plant already treated of (Part I. page 123).

With the exception of the last, there are none of these plants worth culture: they all grow readily in light soil, and are increased with ease by seeds and cuttings.

PASSIFLOREÆ.

PASSIFLORA *adiantifolia*, Adiantum-leaved Passion-flower, B. R. 233, a shrubby creeper, introduced from Norfolk Island in 1792, and flowering from June to August.

P. *incarnata*, Rose-coloured Passion-flower, a shrubby climber, introduced from South America in 1629, and flowering in July and August.

P. *incarnata major*, large Rose-coloured Passion-flower, B. R. 152, a variety of the last species introduced from Brazil.

P. *cærulea racemosa*, Milne's hybrid Passion-flower, a beautiful hybrid (originated at the Fulham Nursery) between P. *cærulea* and P. *racemosa*, the first a hardy and the other a stove species (see Part I. page 35).

P. *edulis*, eatable Passion-flower, B. M. 1989, a climbing shrub, introduced from the West Indies many years ago, and flowering in July and August.

All these plants prefer a loamy soil, and are readily increased by layers. They will also grow, but less easily, by cuttings. They are much esteemed as green-house climbers, especially Milne's hybrid, which in a stove will flower the greater part of the year.

VIOLEÆ.

VIOLA *arborescens*, shrubby Violet, an undershrub, introduced from Spain in 1799, and flowering in April and May. It is a plant of httle beauty, but of easy culture in rich light soil, and propagation by cuttings in sandy loam under a glass.

IONIDIUM *polygalæfolium*, whirl-leaved Ionidium, a perennial, introduced from South America in 1797, and flowering in April and May. It grows in light loam, and is increased by dividing at the roots or by cuttings.

CISTINEÆ.

CISTUS *villosus*, villous Rock-rose, a low frame shrub, introduced from the South of Europe in 1640, and flowering in June and July.

C. *vaginatus*, oblong-leaved Rock-rose, B. R. 225, a low frame shrub, introduced from Teneriffe in 1779, and flowering in April, May and June.

C. *Ledon*, many-flowered Gum-cistus or Rock-rose, a low frame shrub, introduced from the South of France in 1730, and flowering in July and August. It is a most desirable plant, where there is abundance of room, on account of the profusion of large white flowers, with dark spots on the petals, with which it is covered for six weeks together.

C. *ladaniferus*, single-flowered Gum-cistus or Rock-rose, B. M. 112. It is a low frame shrub, introduced from Spain in 1629, and flowering in June and July. It bears a considerable resemblance to the last in its flowers, but has narrower leaves, from which a sort of gummy resin continually exudes. There is a variety C. l. *planifolius*, the flat-leaved Gum Rock-rose.

C. *monspeliensis*, Montpelier Rock-rose, a low frame shrub, introduced from the South of Europe in 1656, and flowering in June and July.

CISTUS *laxus*, waved-leaved Rock-rose, a low frame shrub, introduced from Spain in 1656, and flowering in June and July.

C. *hirsutus*, hairy Rock-rose, a low frame shrub, introduced from Portugal in 1656, and flowering in June and July.

C. *salvifolius*, sage-leaved Rock-rose, a low frame shrub, introduced from the South of Europe in 1548, and flowering in June and July.

C. *heterophyllus*, various-leaved Rock-rose, a low frame shrub, a native of Algiers, and flowering in June and July.

C. *incanus*, hoary-leaved Rock-rose, B.M. 43, a frame shrub, introduced from the South of Europe in 1596, and flowering from June to August.

C. *creticus*, Cretan Rock-rose, a low frame shrub, long since introduced, and flowering in May and June.

C. *albidus*, white-leaved Rock-rose, a low frame shrub, a native of Spain, introduced in 1640, and flowering in June and July.

C. *crispus*, curled-leaved Cistus, a low frame shrub, introduced from Portugal in 1656, and flowering in June and July.

Most of these are beautiful shrubs with showy transient delicate flowers. Some keep them in the open air, and cover them with mats or straw during winter; but to have them flower freely they must have the protection of glass, and, unless they flower freely, of what use is keeping them just within the verge of existence?

HELIANTHEMUM *formosum*, beautiful Sun-rose, B.M. 264, a low frame shrub, a native of Portugal, introduced in 1786, and flowering in May and June.

HELIANTHEMUM *atriplicifolium,* orache-leaved Sun-rose, a low frame shrub, a native of Spain, introduced in 1656, and flowering in June and July.

H. *halimifolium,* sea purslane-leaved Sun-rose, a low frame shrub, introduced from Spain in 1656, and flowering in June and July.

H. *elongatum,* long-peduncled Sun-rose, a low frame shrub, introduced from Spain in 1800, and flowering in July.

H. *algarvense,* Algarvian Sun-rose, B.M. 627, a low frame shrub, a native of Portugal, introduced in 1800, and flowering in July and August.

H. *libanotis,* rosemary-leaved Sun-rose, a low frame shrub, a native of Spain, introduced in 1752, and flowering in June.

H. *umbellatum,* umbel-flowered Sun-rose, a low frame shrub, introduced from the South of Europe in 1731, and flowering from June to August.

H. *lævipes,* cluster-leaved Sun-rose, B. M. 1782, a low frame shrub, introduced from the South of France in 1690, and flowering from June to August.

H. *Fumana,* heath-leaved Sun-rose, a low frame shrub, from the South of France in 1752, and which flowers in June and July.

H. *canum,* hoary Sun-rose, a native of the South of Europe, and a low shrub like all the rest of the species: it was introduced in 1772.

H. *scabrosum,* rough Sun-rose, a low frame shrub, a native of Italy, introduced in 1755, and flowering from June to August.

H. *italicum,* Italian Sun-rose, a low frame shrub, a native of Italy, introduced in 1779, and flowering from July to September.

H. *origanifolium,* marjoram-leaved Sun-rose, a low

(c) 2

frame shrub, a native of Spain, introduced in 1795, and flowering in June and July.

HELIANTHEMUM *Tuberaria*, plantain-leaved Sun-rose, a frame perennial, introduced from the South of Europe in 1752, and flowering in June and July.

H. *canariense*, Canary Sun-rose, a low green-house shrub, introduced from the Canaries in 1790, and flowering in June and July.

H. *glutinosum*, clammy Sun-rose, a low frame shrub, a native of the South of Europe, introduced in 1790, and flowering from May to September.

These plants are of the easiest culture, and the H. *formosum* and a few other species look well when in flower. The flower has this peculiarity, that it only opens during clear sun-shine, and therefore in our climate it often opens and shuts many times a day. The soil which suits them best is a sandy loam with a little peat or leaf-mould: most of them ripen seeds, from which, or from ripened cuttings under a hand-glass, they may be readily increased.

CARYOPHYLLEÆ.

PHARNACEUM *lineare*, linear-leaved Pharnaceum, B. R. 326, a green-house undershrub, introduced from the Cape of Good Hope in 1795, and flowering in May and June.

P. *incanum*, hoary Pharnaceum, B. M. 1883, a green-house undershrub, introduced from the Cape of Good Hope in 1782, and flowering from May to October.

P. *dichotomum*, forked Pharnaceum, a green-house annual, introduced from the Cape of Good Hope in 1783, and flowering in July.

This is a genus of no beauty whatever: it is easily

cultivated in loam and peat, the plants being placed pretty near the light; and propagation is effected by cuttings under a hand-glass in sandy loam.

LINUM *trigynum*, three-styled Flax, B. M. 1100, an evergreen undershrub of very humble growth, introduced from the East Indies in 1799, and producing large yellow flowers from November to January.

L. *narbonense*, Narbonne Flax, B. C. 190, a frame perennial, a native of France, introduced in 1759, and flowering with little beauty in July and August.

L. *suffruticosum*, Spanish Flax, an undershrub of humble growth, but evergreen; a native of Spain, introduced in 1759, and flowering in August.

L. *arboreum*, Tree Flax, B. M. 234, a low single-stemmed shrub, a native of Candia, introduced in 1788, and flowering from May to August.

L. *africanum*, African Flax, B.M. 403, a low shrub, semi-evergreen, a native of the Cape of Good Hope, introduced in 1771, and flowering in June and July.

L. *quadrifolium*, four-leaved Flax, B. M. 431, an evergreen undershrub, a native of the Cape of Good Hope, introduced in 1787, and flowering in May and June.

These plants are of no beauty, though the first species, L. *trigynum*, is of value as furnishing bloom in mid-winter : they all grow readily in loam and peat with a little sand or rotten tan intermixed, and cuttings root readily in sandy loam under a hand-glass. Mr. Sweet remarks in his excellent manual (*The Botanical Cultivator*) that L. *trigynum* is in general much infested with red spiders, but that sprinkling a little flower of sulphur now and then will subdue them (p. 217).

FRANKENIA *Nothria*, Cape Sea-heath, a green-house perennial, a creeper, introduced from the Cape of Good

Hope in 1816, and flowering from June to August. It is of no beauty, but of easy culture in loam and leaf-mould, and cuttings root under a hand-glass in the same soil.

DIANTHUS *japonicus*, Japanese Pink, an evergreen frame perennial, a native of China, introduced in 1804, and flowering from June to October.

D. *carolinianus*, Carolina Pink, a frame perennial, a native of North America, introduced in 1811, and flowering from June to September.

D. *crenatus*, long-cupped Pink, B. R. 256, an ever-green, green-house perennial, a native of the Cape of Good Hope, introduced in 1817, and flowering in August.

D. *arboreus*, shrubby Pink, B. C. 459, an evergreen green-house undershrub, a native of Greece, introduced in 1815, and flowering from June to September.

This genus of plants are of less value in the green-house than the hardy sorts are in the open air, where their glaucous and perpetual green affords a fine clothing for patches and borders in the winter season. D. *arboreus*, however, is a handsome plant when neatly supported by a prop. All of the species grow in any light rich soil, and are propagated by cuttings or pipings like the common pink.

SILENE *fruticosa*, shrubby Catchfly, a frame under-shrub, a native of Sicily, introduced in 1629, and flowering in June and July.

S. *gigantea*, gigantic Catchfly, a green-house biennial, a native of Africa, introduced in 1738, and flowering in June and July.

S. *crassifolia*, thick-leaved Catchfly, a green-house biennial, a native of the Cape of Good Hope, intro-duced in 1774, and flowering in July and August.

SILENE *ornata,* dark-coloured Catchfly, B. M. 382, a green-house biennial, a native of the Cape of Good Hope, introduced in 1775, and flowering from May to September.

S. *undulata,* wave-leaved Catchfly, a green-house biennial, a native of the Cape of Good Hope, introduced in 1755, and flowering in August.

S. *ægyptiaca,* Egyptian Catchfly, a green-house biennial, a native of Egypt, introduced in 1800, and flowering in July and August.

These plants, which are undeserving of culture, excepting as forming part of a botanical collection, thrive well in rich light soil, and cuttings root as readily as those of the common pink or sweet-william.

ARENARIA *procumbens,* procumbent Sandwort, a perennial, introduced from Egypt in 1801, and flowering in July and August. It is a plant of no beauty, but easily cultivated in sandy loam and increased by seeds or cuttings.

LYCHNIS *coronata,* Chinese Lychnis, B. M. 223, a green-house perennial, a native of China, introduced in 1774, and flowering from June to September. It grows in sandy loam and is increased by cuttings like *Dianthus.*

MALVACEÆ.

SIDA *carpinifolia,* hornbeam-leaved Sida, a greenhouse shrub, introduced from the Canaries in 1774, and flowering from July to September.

S. *Sonneratiana,* Sonnerat's Sida, a green-house biennial, introduced from the Cape of Good Hope in 1806, and flowering in June and July.

S. *cresta,* crested Sida, B. M. 330, an annual from Mexico, introduced in 1720, and flowering in June and July.

Sɪᴅᴀ *Dilleniana*, Dillenius's Sida, an annual from Mexico, introduced in 1725, and flowering from July to November.

S. *triloba*, three-lobed Sida, a biennial from the Cape of Good Hope, introduced in 1794, and flowering from July to September.

These plants are of no beauty, but of the easiest culture in any light soil, and they ripen abundance of seeds, from which or from cuttings they may be increased at pleasure.

Mᴀʟᴠᴀ *polystachia*, many-spiked Mallow, a spongy-wooded green-house shrub from Peru, introduced in 1798, and flowering in July and August.

M. *calycina*, large-calyxed Mallow, B. R. 297, a shrub from the Cape of Good Hope, introduced in 1812, and flowering from May to August.

M. *angustifolia*, narrow-leaved Mallow, a green-house shrub, introduced from Mexico in 1780, and flowering in August.

M. *bryonifolia*, bryony-leaved Mallow, a shrub from the Cape of Good Hope in 1731, and flowering in July and August.

M. *asperrima*, roughest Mallow, a shrub from the Cape of Good Hope in 1796, flowering from June to September.

M. *procumbens*, procumbent Mallow, a perennial from South America, introduced in 1815, and flowering from June to September.

M. *abutiloides*, Bahama Mallow, a shrub introduced from the Bahama Islands in 1725, and flowering from June to September.

M. *stricta*, upright Mallow, a shrub introduced from the Cape of Good Hope in 1805, and flowering from May to August.

MALVA *lactea,* panicled white Mallow, a shrub introduced from Mexico in 1780, and valuable for its early
blossoms in January and February.

M. *operculata,* lid-capsuled Mallow, a shrub introduced from Peru in 1795, and flowering in July and
August.

M. *fragrans,* fragrant Mallow, B. R. 296, a shrub
introduced from the Cape in 1759, and flowering from
May to July.

M. *capensis,* Cape Mallow, B. R. 295, a shrub introduced from the Cape of Good Hope in 1713, and valued
on account of its flowers, which appear on the plant every
month in the year.

M. *balsamica,* balsamic Mallow, a shrub introduced
from the Cape of Good Hope in 1800, and flowering
from May to September.

M. *grossularifolia,* gooseberry-leaved Mallow, B. R.
561, a shrub introduced from the Cape in 1732, and
flowering from May to September.

M. *virgata,* twiggy Mallow, a shrub introduced from
the Cape of Good Hope in 1727, and flowering from
May to July.

M. *miniata,* painted Mallow, a shrub introduced from
South America in 1798, and flowering from May to July.

M. *retusa,* blunt-leaved Mallow, a shrub from the
Cape of Good Hope, introduced in 1803, and flowering
from March to May.

M. *tridactyloides,* reflex-flowered Mallow, A. R. 135,
a shrub introduced from the Cape of Good Hope in
1791, and flowering from June to August.

M. *divaricata,* straddling Mallow, A. R. 182, a
green-house shrub, introduced from the Cape of Good
Hope in 1792, and flowering from June to September.

M. *prostrata,* trailing Mallow, a shrub introduced

from the Brazils in 1806, and flowering from June to August.

MALVA *elegans*, elegant Mallow, a shrub introduced from the Cape of Good Hope in 1791, and flowering from May to August.

Some of these plants, especially M. *stricta, capensis* and *grossularifolia*, are free flowerers, and well deserve a place where there is room. They are all of the easiest culture in any light soil: they generally ripen abundance of seeds, and may be increased by that means or by cuttings.

LAVATERA *micans*, glittering Lavatera, a spongy-wooded green-house shrub, introduced from Spain in 1796, and flowering in June and July.

L. *Olbia*, downy-leaved Lavatera, a shrub from the South of France, introduced in 1570, and flowering from June to October.

L. *unguiculata*, clawed Lavatera, a shrub first brought into notice in 1807, and flowering from July to September.

L. *hispida*, hispid Lavatera, a shrub introduced from Algiers in 1804, and flowering in June and July.

L. *triloba*, three-lobed Lavatera, B. M. 2226, a shrub introduced from Spain in 1759, and flowering in June and July.

L. *lusitanica*, Portugal Lavatera, a shrub introduced from Portugal in 1731, and flowering in August and September.

L. *maritima*, sea-side Lavatera, a tree-like shrub, introduced from the South of Europe in 1597, and flowering from April to June.

The last species is an old inmate of the green-house, and valued for its showy red flowers: but most of them require too much room and larger pots than correspond well with a select stock of plants. They grow very freely in any light rich soil, and ripen plenty of seeds, from which, or from ripened cuttings in sand under a bell- or hand-glass, they may be readily propagated.

MALOPE *malacoides*, Barbary Malope, a biennial, a native of Barbary, introduced in 1710, and flowering in June and July. It grows in sandy loam and is increased by seeds or cuttings.

URENA *lobata*, angular-leaved Urena, a shrub from China, introduced in 1731, and flowering in June and July. It grows in loamy soil and may be increased by cuttings. It is however of no beauty.

HIBISCUS *Patersonii*, Norfolk Island Hibiscus, A. R. 286, a spongy-wooded shrub, introduced in 1792, and flowering from June to August.

H. *incanus*, hoary Hibiscus, a perennial introduced from Carolina in 1806, and flowering in September.

H. *militaris*, smooth Hibiscus, a perennial introduced from Louisiana in 1804, and flowering in August and September.

H. *æthiopicus*, dwarf wedge-leaved Hibiscus, a greenhouse shrub, introduced from the Cape of Good Hope in 1774, and flowering in August.

H. *acerifolius*, maple-leaved Hibiscus, a shrub introduced from China in 1798, and flowering from March to June.

H. *speciosus*, superb Hibiscus, B. M. 360, a perennial introduced from Carolina in 1778, and flowering from June to August.

H. *heterophyllus*, various-leaved Hibiscus, B. R. 29, a shrub introduced from New South Wales in 1803, and flowering in August and September.

H. *scaber*, scabrous Hibiscus, a perennial introduced

from Carolina in 1810, which flowers from July to September.

HIBISCUS *pentacarpos*, angular-fruited Hibiscus, a perennial introduced from Venice in 1752, and flowering from July to September.

The green-house species of *Hibiscus* are of no great beauty, and much inferior as ornamental shrubs to the hardy and hot-house kinds. They are however of easy culture in any rich light soil, and they often ripen seeds, from which, or from cuttings in sand under a hand-glass, they may be abundantly increased.

PAVONIA *præmorsa*, bitten-leaved Pavonia, B. M. 436, B. C. 371, a spongy-wooded shrub, introduced from the Cape of Good Hope in 1774, and flowering from June to August. A plant of no beauty, but of easy culture in rich light soil: ripening abundance of seeds, from which, or from cuttings in sand under a hand-glass, it may be readily propagated.

GORDONIA *Lasianthus*, smooth Gordonia, B. M. 668, a shrub, a native of North America, introduced in 1739, and flowering from August to November.

G. *pubescens*, a frame shrub, a native of Carolina, introduced in 1774, and flowering from August to September.

These are very hardy plants and will sometimes bear the winter in the open air, but they never flower well unless in the green-house. They are very ornamental when in flower, and valuable as continuing in blossom till the end of November. The soil they affect is loam and leaf-mould, and they are increased by cuttings, which, as in the case of most of this order of plants, should be ripened before being taken off, and then planted in sand under a glass.

STERCULIACEÆ.

Sterculia *platanifolia*, a tree introduced from China in 1757, and flowering in July. It is of little beauty as a green-house plant, but of easy culture in rich loam, and it is readily increased by ripened cuttings under a hand-glass.

TILIACEÆ.

Mahernia *pinnata*, winged-leaved Mahernia, B.M. 277, an undershrub introduced from the Cape of Good Hope in 1752, and flowering from June to August.

M. *diffusa*, procumbent Mahernia, B. C. 187, an undershrub introduced from the Cape of Good Hope in 1774, and flowering from June to August.

M. *incisa*, cut-leaved Mahernia, B. M. 353, an undershrub introduced from the Cape of Good Hope in 1792, and flowering from July to September.

M. *pulchella*, neat Mahernia, an undershrub introduced from the Cape of Good Hope in 1792, and flowering from July to August.

M. *glabrata*, sweet-scented Mahernia, A. R. 85, an undershrub introduced from the Cape of Good Hope in 1789, and flowering in April and May.

M. *grandiflora*, large-flowered Mahernia, B. R. 224, an undershrub introduced from China in 1791, and flowering from May to August.

These plants are free flowerers, not of bulky or straggling forms, and of very easy culture; but all their flowers are yellow. They come in, excepting M. *glabrata*, at a season when there are abundance of other plants in flower. The two last species are the most appropriate, the first of them as odoriferous, and the other as showy. Loam and leaf-mould, or loam-peat and a little sand, will grow them freely. Cuttings in young wood root under a bell-glass.

SPARRMANNIA *africana*, African Sparrmannia, B. M. 516, a tree introduced from the Cape of Good Hope in 1790, which produces white flowers of no great beauty from March to July. It grows freely in loam and leaf-mould, and cuttings ripened a little root with facility.

GREWIA *occidentalis*, elm-leaved Grewia, B. M. 422, a tree introduced from the Cape of Good Hope in 1690, and flowering from June to September. It is of little beauty, but of easy culture in loamy soil rendered open by sand or a little leaf-mould. Cuttings root in the same soil under a bell-glass.

HERMANNIA *althæifolia*, althæa-leaved Hermannia, B. M. 307, an undershrub introduced from the Cape of Good Hope in 1728, and flowering from March to July.

H. *plicata*, plaited-leaved Hermannia, an undershrub introduced from the Cape of Good Hope in 1774, and flowering in November and December.

H. *candicans*, white Hermannia, an undershrub introduced from the Cape of Good Hope in 1774, and flowering from April to June.

H. *disticha*, round-leaved Hermannia, an undershrub introduced from the Cape of Good Hope in 1789, and flowering from May to August.

H. *salvifolia*, sage-leaved Hermannia, an undershrub introduced from the Cape of Good Hope in 1795, and flowering from April to June.

H. *micans*, glittering Hermannia, an undershrub introduced from China in 1790, and flowering from May to August.

HERMANNIA *involucrata*, involucred Hermannia, an undershrub introduced from China in 1794, and flowering in May and June.

H. *scordifolia*, germander-leaved Hermannia, a low shrub introduced from China in 1794, and flowering from April to November. This species and H. *odorata* are esteemed the most desirable for select collections; the one as in bloom most part of the year, and the other as equally long in bloom and odoriferous.

H. *odorata*, sweet-scented Hermannia, an undershrub introduced from China in 1780, and flowering from February to October.

H. *mollis*, soft-leaved Hermannia, an undershrub introduced from the Cape of Good Hope in 1814, and flowering in May and June.

H. *denudata*, smooth Hermannia, an undershrub introduced from the Cape of Good Hope in 1774, and flowering from May to July.

H. *disermæfolia*, simple-flowered Hermannia, an undershrub introduced from the Cape of Good Hope in 1795, and flowering in March and April.

H. *alnifolia*, alder-leaved Hermannia, B. M. ·299, an undershrub introduced from the Cape of Good Hope in 1728, and flowering from February to May.

H. *cuneifolia*, wedge-leaved Hermannia, an undershrub introduced from the Cape of Good Hope in 1791, and flowering in August and September.

H. *holosericea*, velvet-leaved Hermannia, an undershrub introduced from China in 1792, and flowering in May and June.

H. *hirsuta*, hairy-branched Hermannia, an undershrub introduced from China in 1790, and flowering in May and June.

H. *scabra*, rough-leaved Hermannia, an undershrub

introduced from China in 1789, and flowering in March and April.

HERMANNIA *multiflora*, many-flowered Hermannia, an undershrub introduced from China in 1791, and flowering from March to May.

H. *flammea*, flame-flowered Hermannia, B. M. 1349, an undershrub introduced from China in 1794, and flowering all the year. This is a very desirable species; but is rather less hardy and not such a free flowerer as H. *scordifolia.*

H. *angularis*, angular Hermannia, an undershrub introduced from the Cape of Good Hope in 1791, and flowering in April and May.

H. *hyssopifolia*, hyssop-leaved Hermannia, an undershrub, introduced from China in 1725, and flowering from April to June.

H. *trifurcata*, three-forked Hermannia, an undershrub, introduced from China in 1789, and flowering from April to July.

H. *lavandulifolia*, lavender-leaved Hermannia, B. M. 304, an undershrub, introduced from China in 1732, and flowering from May to September.

H. *filifolia*, thread-leaved Hermannia, an undershrub, a native of the Cape of Good Hope, introduced in 1816, and flowering from May to August.

H. *trifoliata*, three-leaved Hermannia, an undershrub, introduced from China in 1752, and flowering from May to August.

H. *procumbens*, procumbent Hermannia, an undershrub, introduced from China in 1792, and flowering in May and June.

H. *grossularifolia*, gooseberry-leaved Hermannia, an undershrub, introduced from China in 1731, and flowering in April and May.

Hermannia *pulverulenta*, powdered Hermannia, A. R. 161, an undershrub, introduced from China in 1800, and flowering from May to August.

H. *incisa*, cut-leaved Hermannia, an undershrub, introduced from China in 1806, and flowering in June and July.

H. *tenuifolia*, slender-leaved Hermannia, B. M. 1348, an undershrub, flowering in June and July.

These plants are free growers and most of them prolific in flowers, but which unfortunately are yellow. They grow in any rich light soil, and are increased by cuttings in sandy loam. There is a great sameness and very little beauty in their appearance.

SAPINDACEÆ.

Sapindus *marginatus*, marginated Soap-berry, a shrub, introduced from Carolina, and which has not yet flowered in England.

S. *rigidus*, ash-leaved Soap-berry, a shrub, introduced from America in 1759, and flowering from July to September.

These plants are of no beauty; they grow in loamy soil, and cuttings root in sand under a hand-glass.

PITTOSPOREÆ.

Bursaria *spinosa*, thorny Bursaria, B. M. 1767, a shrub, introduced from New South Wales in 1793, and flowering from August to December.

This is an elegant plant and free flowerer; the flowers are white, and though small are very showy from their abundance. It grows in sandy loam, with a little peat or leaf-mould, and cuttings root in sand covered with a gardener's bell.

BILLARDIERA *scandens*, climbing Appleberry, B. M. 801, a green-house climber, introduced from New South Wales in 1790, and flowering from June to August.

B. *mutabilis*, changeable Appleberry, B. M. 1813, a green-house climber, introduced from New South Wales in 1795, and flowering from July to September.

B. *longiflora*, blue-berried Appleberry, B. M. 1507, a climber from Van Dieman's Land, introduced in 1810, and flowering profusely from June to September. It is a very hardy plant, and, probably, in time may be naturalized : it grows rapidly, flowers freely, and is succeeded by dusky blue berries which remain till Christmas.

These plants are all evergreens and climbers, of the easiest culture in light rich loam, and cuttings root in sand; or, as all of them ripen seeds, they may be increased in that way.

PITTOSPORUM *coriaceum*, thick-leaved Pittosporum, A. R. 151, and B. C. 569, a shrub, introduced from Madeira in 1787, and flowering in May: flowers white.

P. *viridiflorum*, green-flowered Pittosporum, B. M. 1684, a shrub, introduced from the Cape of Good Hope in 1806, and flowering in May and June.

P. *Tobira*, glossy-leaved Pittosporum, B. M. 1396, a shrub, introduced from China in 1804, producing its white flowers from March to August: it is a hardy plant, and, being an evergreen, is desirable in either the green-house or conservatory.

P. *undulatum*, wave-leaved Pittosporum, B. R. 16, a shrub, introduced from New South Wales in 1789, and flowering from February to June.

P. *revolutum*, downy-leaved Pittosporum, B. R. 1806, B. C. 506, a shrub, introduced from New South Wales in 1795, and flowering from February to April.

All these plants are evergreens, hardy, and easily cultivated in sandy loam; they are increased by laying, or by grafting on each other, or by cuttings of young wood planted in sand and covered with a bell-glass.

HYPERICINEÆ.

HYPERICUM *balearicum*, warted St. John's Wort, B. M. 137, an undershrub, introduced from Majorca in 1714, and flowering from March to September.

H. *monogynum*, Chinese St. John's Wort, B. M. 334, an undershrub, introduced from China in 1753, and flowering from March to September.

H. *foliosum*, shining St. John's Wort, a low shrub, introduced from the Azore Islands in 1778, and flowering in August.

H. *floribundum*, many-flowered St. John's Wort, a low shrub, a native of Madeira, introduced in 1779, and flowering in August.

H. *canariense*, Canary St. John's Wort, an undershrub, introduced from the Canaries in 1699, and flowering from July to September.

H. *ægyptiacum*, Egyptian St. John's Wort, B.R. 196, an undershrub, introduced from Egypt in 1787, and flowering in June and July.

H. *setosum*, unbranched St. John's Wort, a frame perennial, introduced from Carolina in 1759, and flowering in July and August.

H. *mutilum*, small-flowered St. John's Wort, a frame perennial, introduced from North America in 1759, and flowering from June to September.

H. *nudiflorum*, naked-panicled St. John's Wort, a frame undershrub, introduced from North America in 1811, and flowering in September and October.

HYPERICUM *glaucum*, glaucous St. John's Wort, a frame undershrub, introduced from North America in 1812, and flowering in July and August.

H. *rosmarinifolium*, rosemary-leaved St. John's Wort, a frame undershrub, introduced from Carolina in 1812, and flowering from June to August.

H. *aspalathoides*, aspalathus-like St. John's Wort, a frame undershrub, introduced from Carolina in 1811, and flowering from June to August.

H. *fasciculatum*, clustered St. John's Wort, a frame undershrub, introduced from North America in 1806, and flowering in July.

H. *reflexum*, hanging-leaved St. John's Wort, a green-house undershrub, introduced from Teneriffe in 1778, and flowering from July to September.

H. *maculatum*, spotted St. John's Wort, a frame perennial, introduced from North America in 1789, and flowering in July and August.

H. *crispum*, curl-leaved St. John's Wort, a frame perennial, introduced from Greece in 1688, and flowering in July and August.

H. *undulatum*, wave-leaved St. John's Wort, a frame perennial, a native of Barbary, introduced in 1802, and flowering in July and August.

H. *perfoliatum*, perfoliate St. John's Wort, a frame perennial, a native of Italy, introduced in 1785, and flowering in May and June.

H. *heterophyllum*, various-leaved St. John's Wort, a green-house undershrub, introduced from Persia in 1812, and flowering in July and August.

H. *ciliatum*, fringe-flowered St. John's Wort, a frame perennial, a native of the Levant, introduced in 1739, and flowering in July.

H. *glandulosum*, glandulous St. John's Wort, a green-house undershrub, a native of Madeira, introduced in 1777, and flowering from May to August.

HYPERICUM *tomentosum*, woody St. John's Wort, a frame perennial, a native of the South of Europe, introduced in 1648, and flowering from July to September.

H. *Coris*, heath-leaved St. John's Wort, B. M. 178, a green-house undershrub, introduced from the Levant in 1640, and flowering from May to September.

H. *verticillatum*, whorl-leaved St. John's Wort, a green-house perennial, introduced from the Cape of Good Hope in 1784, and flowering from June to August.

Few of these plants are of any beauty: some of them flower freely, but their flowers are wholly yellow, even to the pistils and stamens. They are all of easy culture in light loam, and are increased by dividing at the root in some cases, by seeds in others, and in all by cuttings in sand, and covered with a crystal bell.

ASCYRUM *pumilum*, dwarf Ascyrum, a perennial, introduced from Georgia in 1806, and flowering from June to August.

A. *Crux Andreæ*, St. Andrew's Cross Ascyrum, a shrub, introduced from North America in 1759, and flowering in July.

A. *hypericoides*, hypericum-like Ascyrum, an undershrub from North America, introduced in 1759, and flowering from July to September.

A. *stans*, large-flowered Ascyrum, an undershrub, a native of North America, introduced in 1806, and flowering from July to August.

These plants, like the Hypericums, are of no beauty, and their culture and propagation are the same as for that genus.

GUTTIFEREÆ.

OCHNA *africana*, purple-flowered Ochna, a shrub, a native of the Cape of Good Hope, introduced in 1816. It has not yet flowered : it grows in sandy peat and loam, and cuttings are rooted in the same soil under a glass.

ELÆOCARPUS *cyaneus*, blue-fruited Elæocarpus, B. M. 1737, a beautiful New Holland shrub, introduced in 1803, and flowering from June to August. It grows well in loam and peat with a little sand, and is increased by ripened cuttings in sand, and covered with a bell-glass.

GERANIACEÆ.

TROPÆOLUM *minus flore pleno*, B. M. 98, double-flowered small Indian Cress, a hardy annual, in its double variety comparatively shrubby and more durable, requiring the protection of the green-house.

T. *majus flore pleno*, B. M. 23, double-flowered large Indian Cress, a plant of a similar character to the preceding. Both are showy inmates of the green-house as to their flowers, which are in perfection from June to October; but their leaves and habits are so salad- and kitchen-garden-like, that we cannot recommend them. They grow in rich light soil, and are increased by cuttings, care being taken that these do not damp off.

T. *peregrinum*, fringe-flowered Indian Cress, B. M. 1351, B. R. 718, an annual, from Peru, introduced in 1775, and flowering from June to October.

T. *pinnatum*, pinnate-flowered Indian Cress, A. R. 535, a perennial, flowering from June to October.

T. *hybridum*, hybrid Indian Cress, a biennial, flowering from June to August.

These plants grow freely in rich light soil, and cuttings root readily in sand under a hand-glass.

OXALIS *monophylla*, simple-leaved Wood-sorrel, a perennial, introduced from the Cape of Good Hope in 1774, and flowering in October and November.

O. *rostrata*, beaked Wood-sorrel, a perennial, introduced from the Cape of Good Hope in 1795, and flowering in October and November.

O. *asinina*, ass's-eared Wood-sorrel, a perennial, introduced from the Cape in 1792, and flowering in November and December.

O. *lancifolia*, spear-leaved Wood-sorrel, a perennial, introduced from the Cape of Good Hope in 1795, and flowering in October and November.

O. *leporina*, hare's-eared Wood-sorrel, a perennial, introduced from the Cape of Good Hope in 1795, and flowering in October and November.

O. *crispa*, curled Wood-sorrel, a perennial, introduced from the Cape of Good Hope in 1793, and flowering in October and November.

O. *fabæfolia*, bean-leaved Wood-sorrel, a perennial, introduced from the Cape of Good Hope in 1794, and flowering in October and November.

O. *laburnifolia*, laburnum-leaved Wood-sorrel, a perennial, introduced from the Cape in 1793, and flowering in September and October.

O. *sanguinea*, blood-leaved Wood-sorrel, a perennial, introduced from the Cape of Good Hope in 1795, and flowering from October to December.

O. *ambigua*, ambiguous Wood-sorrel, a perennial, introduced from the Cape of Good Hope in 1790, and flowering from September to December.

O. *versicolor*, striped-flowered Wood-sorrel, B. M. 155.

O. *pentaphylla*, five-leaved Wood-sorrel, B. M. 1549.

O. *flava*, narrow-leaved Wood-sorrel, B. R. 117.

Besides the preceding there are the following species,

which to those who are curious will form a very unique assemblage of gay little plants.

Oxalis *undulata*, waved-leaved Wood-sorrel
 fuscata, crown-spotted
 tricolor, three-coloured
 variabilis, variable
 grandiflora, great-flowered
 sulphurea, sulphur-coloured
 flaccida, flaccid
 purpurea, purple
 speciosa, specious
 marginata, green-margined
 pulchella, beautiful
 obtusa, blunt-leaved
 lanata, woolly-leaved
 tenella, slender
 compressa, compressed
 sericea, silky
 violacea, violet, B. M. 2215
 caprina, goat's-foot, B. M. 237
 cernua, dropping
 dentata, toothed
 livida, livid
 ciliaris, ciliate-leaved
 arcuata, gland-covered
 linearis, linear-leaved
 cuneifolia, wedge-shaped
 glabra, smooth-leaved

O. *bifida*, cloven-leaved
 filicaulis, bilobed-leaved
 natans, floating
 convexula, convex-leaved
 elongata, elongated
 reclinata, reclining
 polyphylla, many-leaved
 tenuifolia, fine-leaved
 macrostylis, long-styled
 hirta, hairy-stalked
 tubiflora, tube-flowered
 secunda, side-flowering
 multiflora, many-flowered
 rubella, branching red, B. M. 1031
 rosacea, rose-coloured
 repens, creeping-stalked
 reptatrix, creeping-rooted
 incarnata, flesh-coloured
 perennans, perennial
 rubens, red-flowered
 pentaphylla, five-leaved, B. M. 1549
 tomentosa, downy-leaved
 lupinifolia, lupin-leaved
 pectinata, pectinated
 flabellifolia, fan-leaved

This is a genus of pretty little plants, but possessing a great sameness of character, and, excepting one or two of the first described species, of little interest to any but the botanist. Their roots are, generally, bulbs, articulated, jointed, or granulated in a manner peculiar to this genus and one or two others. They grow well in sandy loam, and require only very small pots; great care must be taken not to water them after they have done flowering. They are readily increased by small bulbs or offsets, and many of the species produce seeds. All those species which flower between June and November may be grown in the open air, in the following manner. The bulbs being obtained, plant in beds in the course of the month of April, protect with mats during severe weather, and every night till the end of May; then all covering may be removed for the season. After this, one kind will come into flower after another till October or November; and as each sort finishes flowering, and the leaves begin to fade, take it up and place the bulbs (each sort by itself) in small pots of dry sand, and set them in a dry loft, not colder than a green-house, during winter, till wanted for planting in the open air next spring. Ixias, Gladioluses, and various other green-house bulbs, may be treated in the same manner with complete success. (See Herbert's Essays in the *Botanical Register* and in *Hort. Trans.*)

PHYMATANTHUS *tricolor*, three-coloured Wartflower, the *Geranium bicolor* of gardeners, S. G. 43, B. M. 240, the stem suffruticose and erect with persistent stipules, the leaves various in form, lanceolate, sometimes trifid, deeply toothed and hairy: the flowers with the upper petals a bright scarlet and the lower white. It is a native of the Cape of Good Hope, and flowers freely in equal parts of loam, peat and sand.

P. *elatus*, tall Wartflower, the *Geranium tricolor* of gardeners, S. G. 96, a shrubby branching rugged stem, with lanceolate acute unequally cut leaves, and fine dark red and white flowers marked with black, a handsome free growing plant, flowering the greater part of the year.

MONSONIA *speciosa*, large-flowered Monsonia, S. G. 77, B.M. 73, a suffrutescent stem with few branches, the leaves quinate, smooth, and leaflets bipinnate; the flowers on solitary peduncles, large, yellowish white with a tinge of red, and dark red and black at the centre. It is a most beautiful plant, grows in turfy loam, peat and sand, and is propagated from pieces of the roots planted in pots with their tops above the surface of the mould.

MONSONIA *pilosa*, hairy Monsonia, S. G. 199, a suffruticose branching stem, with palmately five-parted or pinnatifid leaves, and pale cream-coloured flowers tinged with red. A handsome plant which grows in loam and peat, and is increased by cuttings of the shoots or roots placed in a dry airy part of the green-house.

JENKINSONIA *pendula*, pendulous-branched Jenkinsonia, S. G. 188, a pendulous much-branched stem, with bipinnatifid or deeply lacinate leaves, and pinkish purple flowers varied with purple lines. A free grower and flowerer, hardy, and very desirable for training to a trellis.

J. *quinata*, quinate-leaved Jenkinsonia, S. G. 79, B. M. 547, a shrubby flexuose stem, the branches covered with a powdery pubescence, the leaves alternate and five-parted, and the flowers yellowish white marked with red. It is a handsome, rare, and curious plant, requiring the warmest part of the green-house; it grows in loam, peat and sand, and is increased by cuttings in the same soil.

J. *tetragona*, square-stalked Jenkinsonia, S. G. 99, B. M. 136, a shrubby succulent much-branched stem, with fleshy cordate leaves and long-petalled red and white flowers. There is a variegated variety, and both require support or training against a wall.

OTIDIA *carnosa*, fleshy-stalked Otidia, S. G. 98, a shrubby succulent branching stem with leaves sometimes ternate or pinnatifid, and small white flowers. Being of a succulent nature, it requires a dry airy situation to keep it in good health.

OTIDIA *laxa*, loose-panicled Otidia, S.G.196, a shrubby succulent stem, with smooth pinnate or pinnatifid leaves and white flowers marked with pale blue. Introduced in 1821 from the Cape of Good Hope, and grows in loam, peat and sand, with the pots well drained.

ISOPETALUM *Cotyledonis*, hollyhock-leaved Isopetalum, S. G. 126, a succulent thick shrubby branching stem, with peltate, cordate, veined rugose leaves and fine white flowers. It requires the same treatment as the tuberous *Pelargoneæ*.

CAMPYLIA *carinata*, keeled-stipuled Campylia, S. G. 21 and 43, a suffruticose stem, requiring support, with oval smooth toothed leaves, and flowers pale below and very dark in the upper petals. The plant rather tender.

C. *holosericea*, silky-leaved Campylia, S. G. 75, a shrubby erect rugged stem, leaves roundly ovate, obtuse and very downy, and the flowers pale and dark red and black. It is a new hybrid and one of the easiest managed of the genus. Though its anthers are always without pollen, yet it may set freely if set with the pollen of any of the same tribe, and hence an endless source of new sorts.

C. *blattaria*, downy-leaved Campylia, S. G. 88, a shrubby branching stem, with roundly oval plaited leaves and purple and white flowers. It flowers from an early period in spring, and continues in bloom till late in autumn.

C. *cana*, a shrubby erect branching stem, with roundly ovate obtuse leaves and pale red flowers. One of the largest-flowered species of the genus.

C. *verbasciflora*, mullein-leaved Campylia, S. G. 157, a suffruticose stem, with roundly oval obtuse leaves and pale red flowers. A handsome hybrid of the usual culture.

CICONIUM *reticulatum*, netted-veined Ciconium, S. G. 143, a shrubby stem branching but little, the leaves cordately uniform, obtusely five-lobed and strongly marked with a zone; the flowers of a fine deep scarlet veined with a darker colour. It is a hybrid from the common horse-shoe Geranium *(Ciconium zonale)*, and probably C. *Fothergilli*, and is as hardy and prolific in flower as its parents.

C. *cerinum*, wax-flowered Ciconium, S. G. 176, a shrubby erect stem with roundly uniform leaves and deep red and paler flowers, having an airy appearance. It is of a succulent nature and must not be over-watered: in other respects it is quite hardy, and readily strikes from cuttings placed in any sheltered situation.

ERODIUM *hymenodes*, three-leaved Heron's-bill, S. G. 23, B. M. 1174, a perennial with numerous stalks covered with soft hairs; the leaves opposite ternate or deeply three-lobed; the flowers of a pale red streaked with fine red lines. It is a native of Barbary, and was introduced in 1789.

E. *incarnatum*, flesh-coloured Heron's-bill, S. G. 94, B. M. 261, a suffruticose stem with few branches, and rough dark green glossy cordate leaves; the flowers flesh-coloured with a circle of deeper red and white near the centre. A handsome and curious plant introduced from the Cape of Good Hope in 1787.

E. *crassifolium*, hoary-leaved Heron's-bill, S. G. 111, a suffruticose branching stem, with pinnatifid or deeply lacinated leaves, and pale purple or reddish flowers. It is a native of Cyprus and Barbary, and now rather rare in collections.

DIMACRIA *sulphurea*, sulphur-coloured Dimacria, S. G. 163, a tuberous branching root, stemless, but with erect pinnate carrot-like leaves and a head of sulphur-coloured flowers. A hybrid of free growth, which flowers abundantly and ripens seeds.

DIMACRIA *pinnata*, wing-leaved Dimacria, S. G. 46, B. M. 579, a tuberous-rooted stemless perennial with pinnate leaves, and pale flowers streaked with red. It is a curious plant, and requires to be grown in small pots, well drained, and in equal parts of turfy loam, peat and sand.

D. *astragalifolia*, astragalus-leaved Dimacria, a tuberous root crowned with clusters of leaves, ternate or unequally pinnate, the flowers with long narrow yellowish petals marked with red. A handsome plant, flowering all the summer and bearing plenty of seeds.

D. *bipartita*, forked-leaved Dimacria, S. G. 142, a tuberous root, stemless, but crowned with spreading, pinnate, hairy, light green leaves; the flowers a pale yellow or cream colour marked with red. A hybrid between a species of this genus and one of *Hoarea*, a free grower and abundant flowerer.

GERANIUM *Nepalense*, Nepal Crane's-bill, S. G. 12, a herbaceous perennial with procumbent spreading branchy stems, cordate, five-lobed leaves, and small dark red flowers. Introduced from Nepal in 1819.

G. *Wallichianum*, Wallich's Crane's-bill, S. G. 90, a perennial with spreading branches, trifid reticulately veined leaves, and fine deep red striated flowers. The plant introduced from Nepal, hardy and probably fit to stand our winters.

G. *pilosum*, pointed-anthered Crane's-bill, S. G. 119, a perennial with forked stems, opposite uniform leaves, and small pale red flowers marked with a circle of white at the base of the petals. It is a native of New Zealand, introduced in 1820, and may possibly turn out hardy enough to stand our winters.

GRIELUM *tenuifolium*, slender-leaved Grielum, S. G.

171, a perennial succulent root with bipinnatifid leaves
and large yellow flowers. A beautiful and rare plant,
rather difficult to preserve through the winter; it dislikes
both moisture and heat, and prefers a dry airy part of
the green-house and light loose soil. At the Cape of
Good Hope it grows in sandy or gravelly situations.

HOAREA *corydaliflora,* fumitory-flowered Hoarea,
S. G. 18, a stemless perennial, with a tuberous root
surmounted by a cluster of pinnate hairy leaves, the
flowers small and of a pale yellow colour.

H. *setosa,* bristle-pointed Hoarea, S. G. 38, a tuberous-rooted perennial, the tubers finger-shaped and
producing other small tubers; leaves in clusters on the
crowns of the tubers, generally much divided; flower
small and of a pale red. When the roots of this plant are
in a dormant state, they should be kept without water.

H. *atra,* dark brown Hoarea, S. G. 72, a tuberous
turnip-shaped root with a scaly bark; the leaves in
clusters at the crown of the root, very variable, entire,
ternate, or pinnatifid; the flowers small and of a black
purple colour. It flowers in autumn, and is best increased
by seeds.

H. *melanantha,* black-flowered Hoarea, S. G. 73, a
scaly fusiform tuberous root, under a crown of dark
green, variable, much cut leaves; the flowers in an umbel,
small and very black. It is a curious plant, and seeds
freely.

H. *reticulata,* netted-petalled Hoarea, S. G. 91, roots
tuberous like those of a turnip radish, crowned with
oblong elliptically lanceolate leaves, from which proceed
stems producing heads of small striated red and white
flowers. A pretty little plant lately introduced from the
Cape, and requiring attention as to watering like most
other tuberous plants of this order.

HOAREA *ovalifolia*, oval-leaved Hoarea, S. G. 106, a tuberous root crowned with radiated leaves, widely oval, obtuse, and entire, with white flowers consisting of long narrow petals.

H. *violæflora*, white violet-flowered Hoarea, S. G. 123, a tuberous root, with a short stiff scaly stem, with pinnatifid or ternate leaves and small white flowers.

H. *elegans*, elegant Hoarea, S. G. 132, a tuberous root with no stem, but crowned with variable pinnatifid hoary leaves; the flowers small, of a deep scarlet and white.

H. *nutans*, nodding-flowered Hoarea, S. G. 135, a tuberous carrot-like root with carrot-like leaves and pale yellow flowers; one of the hardiest species of the genus and producing ripened seeds.

H. *atro-sanguinea*, dark crimson-coloured Hoarea, S. G. 151, a tuberous root without a stem; the leaves crowning the root in a radiate form, pinnatifid and hairy; the flowers small and of a dark red colour. It is a hybrid between a *Hoarea* and *Dimacria*.

H. *selinifolia*, milk parsley-leaved Hoarea, S. G. 159, a tuberous root branching out into small tubers, the leaves pinnate, very hairy, and light green, and flowers deep purple. A neat little mule production.

H. *varia*, various-leaved Hoarea, S. G. 166, a tuberous root stemless with variable leaves, simple, ternate or pinnatifid, and dark purple flowers. A hybrid of the usual culture.

H. *radicata*, fleshy fringed-leaved Hoarea, S.G. 174, B. M. 1718, a tuberous root, stemless, with oblong elliptic entire hairy leaves and yellow flowers. An old inhabitant of the green-house: kept in a warm situation it begins to grow about Christmas, when it should be repotted in fresh soil and a little water given it; and as it advances it must be watered regularly, when quite dry, but never over the leaves.

HOAREA *nivea*, snowy white Hoarea, S. G. 182, a tuberous root without a stem; the leaves smooth, shining, dark green, simple, ovate, lanceolate, slightly fringed; the flowers small and of a snowy whiteness. A small plant, but considered a genuine species.

PELARGONIUM *striatum*, streak-flowered Stork's-bill, or Davey's Fairy Queen Geranium, S. G. 1, a spongy shrub, with streaked flowers, which appear early in spring.

P. *ignescens*, fiery-flowered Stork's-bill, S. G. 2, B. C. 109, a shrubby stem, with cordate leaves and scarlet and black flowers in abundance during the greater part of the year.

P. *blandum*, blueish-flowered Stork's-bill, S. G. 4, a shrubby stem with cordate five-lobed leaves, and light flowers, appearing in succession from April to September. It is known in some nurseries as the Waterloo, and in others as the Diana Geranium.

P. *melessinum*, balm-scented Stork's-bill, S. G. 5, a shrubby branching stem, with deeply five-lobed leaves, and dark red and black flowers in abundance. It is a free grower, and takes up but little room : its leaves when rubbed have the exact fragrance of common balm.

P. *Vandesiæ*, Comtesse de Vandes' Stork's-bill, S. G. 7, a shrubby erect stem, not much branched but pubescent, with leaves palmatifid and deeply divided, and light red flowers. It is a hybrid, raised by Sir R. Hoare.

P. *obtusilobum*, blunt-lobed Stork's-bill, S. G. 8, a shrubby branching stem with deeply three-lobed leaves, and dark red flowers which appear abundantly.

P. *pannifolium*, cloth-leaved Stork's-bill, S. G. 9, a

shrubby upright stem, the branches covered with soft villous hairs, the leaves cordate and slightly lobed, and the flowers white streaked with red. It is a hybrid, an abundant flowerer, and continues in bloom till late in autumn.

PELARGONIUM *Mostynæ*, Mrs. Mostyn's Stork's-bill, S. G. 10, a shrubby upright stem with few branches; the leaves cuneiform and pubescent on both sides, and the flowers of a fine red. It is a very hardy plant, and an early flowerer.

P. *pustulosum*, blistered-leaved Stork's-bill, S.G.11, a shrubby stem very much branched, the leaves deeply three-lobed, and the flowers small and pale. It continues flowering till late in autumn, and has a citron-like scent.

P. *monstrum*, cluster-leaved Stork's-bill, S. G. 13, a shrubby thick and irregularly swollen stem, rather succulent, and producing many short thick branches with crowded leaves of a round kidney shape, and very dark green. The flowers are in close heads and light red, but they appear but seldom.

P. *cortusæfolium*, cortusa-leaved Stork's-bill, S. G. 14, a shrubby succulent very rough stem, short branches, cordate lobed leaves, and pale flowers. Being of a succulent nature it requires a sandy soil and little water.

P. *carduifolium*, cockle-shell-leaved Stork's-bill, S. G. 15, a shrubby branching stem growing to a great size, with large rigid leaves wedge-shaped at the base, and large red and black flowers. It is a hardy plant and very showy.

P. *lineatum*, striped-flowered Stork's-bill, S. G. 16, a shrubby stem much branched, the branches short and crowded, the leaves two-ranked, small, cordate and cuneate, and the flowers red and dark red. It forms a neat little bush, and flowers late in the autumn.

PELARGONIUM *multinerve*, many-nerved Stork's-bill, S. G. 17, a shrubby upright stem not much-branched, the leaves of a roundish kidney-shape, and flowers of a deep red. It is very hardy, and continues in bloom till late in autumn.

P. *dumosum*, bushy Stork's-bill, S. G. 19, a shrubby much-branched stem, with numerous small reniform obtuse leaves, and white and dark purple flowers. A compact little bush, flowering till late in autumn.

P. *Dennisianum*, Dennis's Stork's-bill, S. G. 20, a shrubby branching stem, with large reniform or cordate five-lobed leaves, and fine large dark red-striped flowers. It is a handsome plant, of easy culture, a free flowerer, and continues in bloom till late in autumn.

P. *adulterinum*, hoary trifid-leaved Stork's-bill or Kutusoff's Geranium, S. G. 22, a shrubby stem, of a woody texture, very much-branched, the leaves cordate and trifid, and the flowers of a fine dark red. It is one of the earliest flowerers of the genus.

P. *solubile*, dissolvible-coloured Stork's-bill, or Duchess of Gloucester's Geranium, S. G. 24, a shrubby stem with few branches, the leaves kidney-shaped, and the flowers large, and of a fine deep red mixed with darker red. It is a curious circumstance of this plant, that water dissolves the colour of its petals.

P. *obtusifolium*, blunt-leaved Stork's-bill, S. G. 25, a shrubby stem, of a woody texture, and much branched; the leaves flat, an inch long and two inches wide, deeply three-lobed, the flowers of a pale and darker red. A free grower and abundant flowerer.

P. *eximium*, select Stork's-bill, S. G. 26, a shrubby

erect stem with hairy branches, cordate, undulate, rugged leaves, and fine large flowers, pale red, dark red, and black. It is a strong grower, and requires a rich loamy soil.

PELARGONIUM *papilionaceum*, butterfly Stork's-bill, S. G. 27, an erect branching shrubby stem; the branches terminated with large panicles of flowers, red and dark; the leaves are roundly cordate and bluntly crenate. It is a handsome vigorous plant, which has been long in cultivation.

P. *scintillans*, sparkling Stork's-bill, S. G. 28, a shrubby much-branched stem with flexuose purplish branches, wedge-shaped five-lobed leaves, and deep scarlet flowers. The plant rather delicate.

P. *grandiflorum*, great-flowered Stork's-bill, S. G. 29, a shrubby erect stem not much branched, with smooth glaucous leaves deeply five- to seven-lobed, and large white flowers. It is rather scarce at present, and is also tenderer than any of the other species.

P. *rubescens*, the Countess of Liverpool's Stork's-bill, S. G. 30, the stem shrubby, tall, and erect, and not much branched; the leaves cordate, five-lobed, and undulate; and the flowers large, of a fine dark red above and the lower petals paler. It is a free strong-growing plant.

P. *pulchellum*, nonesuch Stork's-bill, S. G. 31, B. M. 524, a suffruticose short succulent stem, with oblong sinuate leaves, and pale whitish flowers with streaks of red. It is a native of the Cape of Good Hope, and was introduced in 1795; it flowers from March to May, and must be but sparingly watered after the bloom is over.

P. *Daveyanum*, Davey's Stork's-bill, S. G. 32, the stem shrubby, branching, and hairy; the leaves either cordate or kidney-shaped, slightly five-lobed; the flowers of a fine dark scarlet above and the lower petals lighter. It is a hybrid, by Davey of the King's-road, a free grower, abundant flowerer, and continues in bloom a long time.

PELARGONIUM *involucratum*, large-bracted Stork's-bill, S. G. 33, the stem, shrubby, erect, growing to a great size, and not much branched; the leaves are kidney-shaped or cordate; and the flowers very large, white, streaked with dark purple. It is of hybrid origin, and there are several varieties, all "fine showy plants and free growers, thriving well in a light turfy loam, mixed with a little peat or leaf-mould."

P. *coriandrifolium*, Coriander-leaved Stork's-bill, S. G. 34, a herbaceous flexuose branching stem, with bipinnate, smooth, and shining leaves, and white flowers streaked with red. It is a curious plant, a free grower, and easily cultivated.

P. *oblatum*, oblate-leaved Stork's-bill, S. G. 35, a shrubby branching stem, growing to a great size; the leaves very large, dark green, five or six inches long and six inches wide, cordate five- to seven-lobed, and imbricate at the base; flowers large and light and dark red finely marked. It is a fine showy plant, hardy and easily managed.

P. *elegans*, elegant Stork's-bill, S. G. 36, a suffruticose, erect, and rugged stem, owing to the persistent stipules; the leaves rigid and nearly orbicular, and the flowers white and finely streaked with purple.

P. *Seymouriæ*, Mrs. Seymour's Stork's-bill, S. G. 37, a shrubby branching hairy stem, with cordate leaves, and dark red flowers, marked with black. It is a hardy free growing plant of hybrid origin.

P. *ornatum*, ornate Stork's-bill, S. G. 39, a shrubby stem, much-branched; the leaves small, wider than long, between cordate and wedge-shaped, the flowers with a blotch of pink in the upper petals.

PELARGONIUM *pavonuim*, peacock-spotted Stork's-bill, S. G. 40, the stem shrubby and branching; the leaves wedge-shaped at the base, inclining to cordate; the flowers of a fine bright red above, and the lower petals of a pale scarlet. It is a hybrid, raised by Mr. Colville of the King's-road, a free grower, and continues in blossom all the summer and till late in autumn.

P. *floridum*, abundant-flowering Stork's-bill, S. G. 41, a shrubby stem very much branched; the leaves cordate and a little undulate; and the flowers pale red, dark red, and black. It is a free-growing hybrid, very hardy, and blooming till late in the autumn.

P. *humifusum*, trailing Stork's-bill, S. G. 42, the stems herbaceous, prostrate, spreading flat on the ground; the leaves cordate, and three-parted or five-lobed; the flowers very small, of a pale red, and few together. It seems to be a native of the Canaries, and flowers all the summer in any rich light soil.

P. *Lousadianum*, Miss Lousada's Stork's-bill, S. G. 44, a shrubby erect branching stem, three-lobed leaves, and fine large pale and dark red flowers. It is a hardy free-growing plant, seldom rising above eighteen inches.

P. *ardens*, glowing Stork's-bill, S. G. 45, B. M. 139, a suffruticose thick little branched stem, with leaves scarcely two alike, cordate, oblong, or ternate, and small very dark scarlet and black flowers. It is a handsome plant, much admired for the brilliancy of its flowers, and their appearance during the greater part of the year.

P. *nervosum*, prominent-nerved Stork's-bill, S. G. 47, a shrubby branching stem, kidney-shaped five-lobed leaves, and fine large red and black flowers. It is a fine

flowering hybrid, and used to be called Princess Charlotte's Geranium.

PELARGONIUM *reniforme,* kidney-leaved Stork's-bill,
S. G. 48, B. M. 493, a shrubby stem with succulent
branches, small kidney-shaped leaves, and deep red
flowers without white. It is a pretty plant, and delights
in sandy loam, but must not have much water when in
a dormant state.

P. *particeps,* participant Stork's-bill, S. G. 49, a
shrubby erect branching stem, with cordate five-lobed
leaves, and deep red flowers without white. It is a cu·
rious mule, and valued as flowering late in autumn, when
most of the more showy kinds are over.

P. *Boyleæ,* the Countess of Cork's Stork's-bill, S. G. 50,
a shrubby much-branched stem, with distichous cuneate
leaves, and fine large flowers, white, and variegated with
red and dark purple. It is a free grower, and flowers
early in spring and till the end of summer.

P. *lobatum,* cow-parsnip-leaved Stork's-bill, S. G. 51,
B. M. 1986, a large rough tuberous root, crowned with
large various-formed leaves ; the flowers small, of a very
dark purple, fragrant during the night, but without scent
during the day. This is a beautiful and curious plant.
Mr. Sweet in his *Geraniaceæ* says, " We have seldom observed it in flower, which is probably owing to the mode
of treatment. When the plants are in a dormant state, they
require scarcely any water till they show an inclination
to grow ; they should then be fresh potted in an equal
mixture of turfy loam, peat, and sand, with the pots well
drained ; they must then be regularly supplied with water ; as they grow they may be shifted into larger-sized
pots if required, and they will flower as freely as any
other kind : the best method of propagating them is by

the little tubers of the roots, which may be planted several in one pot in the same kind of soil: their tops
must be just above the surface of the mould, and they
require scarcely any water till the wound is callused
over, when they may be watered regularly, and they will
soon make durable young plants." *Geran.* 51.

PELARGONIUM *pinguifolium*, greasy-leaved Stork's-
bill, S. G. 52, the stem shrubby, much branched, the
branches succulent; the leaves between kidney- and
heart-shaped, and the flowers large, pale, and dark red
veined. The plant is rather tender, and must not be over-
watered.

P. *opulifolium*, Guelder-rose-leaved Stork's-bill, S. G.
53, a shrubby erect stem with few branches, and cordate, wide, concave, five-lobed leaves; the flowers very
large, of a deep red, finely streaked with a darker red.
It is a free-growing plant, thriving in turfy loam and
peat.

P. *echinatum*, prickly-stalked Stork's-bill, S. G. 54,
B. M. 309, the root fleshy and producing tubers; the
stem shrubby and succulent; the leaves cordate, from
three- to five-lobed, and the flowers white with bright
red spots.

" This handsome species is one of the most desirable
plants of the genus for any collection, as it begins to
flower in autumn after most of the other kinds have done,
and if kept in a warm situation it will continue to bloom
till late in the spring. As soon as it has done flowering
and becomes dormant, it should be watered very sparingly, requiring very little till it begins to grow again; it
may then be shifted, and the greater part of the mould
taken from its roots; then planted in a fresh pot, in an
equal mixture of loam, peat, and sand, and regularly
watered; and as it increases in size, requires changing

into larger pots, for this is the only way to flower it in perfection. A warm part of the green-house, or the coolest and most airy part of the hot-house, is the most suitable for it in winter. Cuttings root readily if planted when in a growing state, and set in the green-house." Sweet's *Geran.* 54.

PELARGONIUM *ignescens* var. *sterile,* barren-anthered flame-flowered Stork's-bill, S. G. 55, B. C. 109, a shrubby much-branched stem, with leaves trifid or deeply three-lobed, and flowers of a deep scarlet with black spots. It is much esteemed on account of the brilliant colour of the flowers.

P. *inodorum,* scentless Stork's-bill, S. G. 56, a herbaceous, much-branched, spreading stem, with small opposite cordate leaves, and small heads of red and white flowers. It is a curious little plant, but not to be recommended for ornamental culture.

P. *glaucum,* glaucous-leaved Stork's-bill, S. G. 57, B. M. 56, a shrubby erect stem with few branches; lanceolate or spoon-shaped, entire, smooth, and glaucous leaves; and pale yellow and white flowers spotted with red. It is a beautiful and singular plant, somewhat tender, and apt to suffer in winter from wet.

P. *sæpeflorens,* frequent-flowering Stork's-bill, S. G. 58, a shrubby brown stem, with flat cordate five-lobed leaves, and red flowers with darker red and black spots. It is a desirable hybrid, as being in flower the greater part of the year. It begins to bloom, Sweet observes, at the end of summer, " and if kept in a warmish situation it will continue to flower through the winter till late in spring."

P. *bellulum,* neat Stork's-bill, S. G. 60, a shrubby dwarf stem, with five-lobed wedge-shaped leaves, and dark red flowers finely veined with darker red, purple, and black. It forms a pretty little bush, and is profusely covered with flowers all the summer.

PELARGONIUM *gibbosum*, knotted Stork's-bill, S. G. 61, a shrubby succulent stem swollen at the joints, the branches erect or slightly spreading; the leaves ternate, smooth, and glaucous, and the flowers of a greenish yellow. It is a curious plant, esteemed for the pleasant odour it diffuses in the evening when in bloom, though it is quite scentless during the day.

P. *optabile*, desirable Stork's-bill, S. G. 62, a shrubby much-branched stem, with rough, cordate, five-lobed leaves, and large flowers, white blotched with dark purple. It is a hardy plant, and makes a compact little bush eighteen inches high.

P. *hybridum*, mule Stork's-bill, S. G. 63, a shrubby stem, with numerous short flexuose branches, round kidney-shaped leaves, and deep scarlet flowers with dark lines and without white. It is considered an original species, notwithstanding the specific name: it is rather tender, and requires a soil of turfy loam and leaf-mould, and to be not over-watered in winter.

P. *Breesianum*, Breese's Stork's-bill, S. G. 64, a shrubby branching stem, with cordate, oval, slightly-lobed leaves, and fine deep red flowers varied with white and dark red spots and lines. It is a hybrid, hardy, and flowering freely great part of the year.

P. *imbricatum*, imbricate-petaled Stork's-bill, S. G. 65, a shrubby flexuose stem, irregularly swollen at the joints; the leaves nearly trifid, curled, and plaited; the flowers in large heads, the petals obtuse, white round the edges, aud dark red in the middle. It is a fine strong-growing plant, and produces abundance of flowers, through the summer, till late in the autumn.

P. *pectinifolium*, scallop-shell-leaved Stork's-bill, S. G.

66, a shrubby smooth stem much branched, with smooth kidney-shaped plicate furrowed leaves, and fine large dark and light red flowers. A free-growing plant, and an abundant flowerer.

PELARGONIUM *cordatum*, heart-leaved Stork's-bill, S. G. 67, B. M. 165, a shrubby erect stem with few branches; the leaves flat, cordate, acute, and dentate, and the flowers large, purplish red with dark streaks. It is a fine strong-growing plant, introduced from the Cape of Good Hope in 1774, and flowering from March to October.

P. *australe*, New Holland Stork's-bill, S. G. 68, a short suffruticose stem with numerous hairy branches, and cordate leaves with large unequal crenatures; the flowers white and bright red in streaks. It is a neat little plant, introduced in 1792 from New South Wales, and flowering and ripening seeds abundantly.

P. *fulgidum*, celandine-leaved Stork's-bill, S. G. 69, a shrubby flexuose stem with few branches; the leaves ternate, and leaflets sessile, deeply toothed, and pinnatifid; the flowers small, of a deep scarlet with dark spots and lines. The flower is one of the most brilliant of the genus: the plant, being rather succulent, requires attention as to water.

P. *coarctatum*, close-leaved Stork's-bill, or Lavalette's Geranium, S. G. 70, a shrubby much-branched stem, with numerous crowded cordate or kidney-shaped leaves, and large deep red flowers streaked with black. A pretty little hybrid, and hardy.

P. *mixtum*, mixed Stork's-bill, S. G. 71, a shrubby erect branching stem, with cordate, nearly distichous leaves, of an agreeable scent, and white flowers marked with pale red. A neat little bushy plant, flowering from the early part of summer till late in autumn.

PELARGONIUM *Thynneæ*, the Marchioness of Bath's Stork's-bill, S. G. 74, a shrubby much-branched stem, with short crowded branches and flat wide kidney-shaped leaves, and large flowers, deep red and black above, and pale red and white below. A strong-growing plant, and an abundant flowerer, the flowers large in proportion to the size of the plant.

P. *sanguineum*, crimson Stork's-bill, S. G. 76, a shrubby succulent stem, not much branched, the branches knotted at the joints and glaucous; the leaves decompound, smooth, and of a light green; and the flowers of a deep scarlet or blood colour marked with black. It is a new introduction from Germany, and seems to grow luxuriantly.

P. *versicolor*, various-coloured Stork's-bill, S. G. 78, a shrubby branching stem, with flat cordate five-lobed leaves, and large dark and light red flowers streaked with red and black. A free grower, and abundant flowerer.

P. *Hoareanum*, Hoare's Fair Rosamond Stork's-bill, or Fair Rosamond Geranium, S. G. 80, the stem shrubby, of a woody texture; the branches numerous, rigid, flexuose, and clothed with small dark green cordately ovate leaves; the flowers large, not many together, white, dark red, and black. It is a hybrid, and one of the handsomest of the genus. Cuttings must be taken off in very young wood, the old taking a long time to make roots.

P. *calycinum*, large-calyxed Stork's-bill, Brown's Countess of Roden, or Rose-brilliant Geranium, S. G. 81, a shrubby much-branched stem, with leaves orbicularly reniform, scarcely lobed and hairy; the flowers of a deep red, and finely veined and marked with red and black. There are three or more varieties, all free growers and flowerers.

PELARGONIUM *atrofuscum*, dark–brown–marked Stork's-bill, S. G. 82, a shrubby branching stem clothed with long hairs; leaves wedge-shaped, with short foot-stalks; flowers of a dark or purplish brown, marked with still darker brown. It is a handsome plant of hybrid origin, a free grower and abundant flowerer.

P. *macranthon*, large-flowered Stork's-bill, S. G. 83, a shrubby flexuose stem, with leaves inclining to be succulent, hairy on both sides, reniform and deeply toothed; the flowers very large, white, finely streaked, and blotched with purple. It is one of the largest-flowered hybrids that has yet been originated.

P. *filipendulifolium*, dropwort-leaved Stork's-bill, S. G. 85, B. 1641, subcaulescent, with a large brown scaly tuberous root; the leaves pinnatifid or laciniate, light-green inclining to glaucous; the flowers small, greenish, white, and purple. It grows in turfy loam, peat, and sand, and is increased by tubers of the root.

P. *Colvillii*, Colville's Stork's-bill, S. G. 86, a shrubby erect stem with hairy branches, rigid, cordate, undulate leaves, and deep scarlet and black flowers. A fine hybrid production, originated in Colville's nursery in 1820.

P. *Baileyanum*, Bailey's Stork's-bill, S. G. 87, a shrubby branching stem, with kidney-shaped truncate leaves, and large white and dark brown flowers. It is quite hardy, and a free flowerer.

P. *obscurum*, darkened-petaled Stork's-bill, S. G. 89, a shrubby much branched stem, with oblately cordate leaves, and fine dark red, purple and white flowers. One of the finest hybrids yet originated, hardy, and a free flowerer.

PELARGONIUM *Husseyanum*, Lady Mary Hussey's Stork's-bill, or Brown's Duke of York Geranium, S. G. 92, a shrubby branching stem, with dark green cordate lobed leaves, and fine large dark red flowers. A strong and robust hybrid, flowering abundantly all the summer.

P. *incisum*, cut-leaved Stork's-bill, S. G. 93, a shrubby branching stem, with ternate dark green pubescent leaves, and small narrow-petaled white flowers striated with red lines. A rather tender plant, requiring an airy situation and care as to water.

P. *scutatum*, shield-leaved Stork's-bill, S. G. 95, a shrubby much-branched stem, with peltate five-lobed leaves, and large white flowers finely marked with red. A handsome plant, raised in 1819 from seed received from the Cape of Good Hope.

P. *bicolor*, two-coloured Stork's-bill, S. G. 97, B. M. 201, a shrubby branching knotted stem, with cordate trifid undulate leaves, and deep dark brown and red flowers. Introduced in 1778, since which two varieties of it have been procured from seeds.

P. *villosum*, villous Stork's-bill, or Wood's Geranium Majestum, S. G. 100, a shrubby branching stem, with roundish truncate leaves, and very large red flowers marked with dark purple. A fine plant, of hybrid origin, requiring a dry airy situation in the green-house.

P. *Blandfordianum*, hoary-leaved Stork's-bill, S. G. 101, a shrubby gouty-jointed branching stem, with flat palmately seven-lobed leaves, and small white flowers marked with red. A curious free-growing hybrid, very hardy.

P. *nanum*, pygmy Stork's-bill, S. G. 102, a short suffruticose stem, with cordate three-lobed leaves, and deep scarlet flowers with dark lines. A curious and handsome little hybrid, slow of increase, because it produces few shoots.

PELARGONIUM *Lamberti*, Lambert's Stork's-bill, S. G. 104, a shrubby much-branched stem, with wedge-shaped trifid leaves, and fine red flowers marked with darker red and brown. A compact bush, flowering abundantly for the greater part of the year.

P. *difforme*, various-leaved Stork's-bill, or Davey's Princess Augusta Geranium, S. G. 105, a shrubby much-branched stem, with rigid, deformed, cordate, or cuneate leaves, and large-petaled flowers streaked and blotched with dark red and brown. It is a hardy and easily managed hybrid, requiring very little water in winter.

P. *pulchrum*, gay Stork's-bill, S. G. 107, a shrubby branching stem, with kidney-shaped slightly lobed leaves, and dark red and white flowers veined with purple and brownish red.

P. *concinnum*, comely Stork's-bill, S. G. 108, a shrubby branching stem, with truncate, trifid, undulate leaves, and fine scarlet flowers marked with a darker streak of the same colour and with black lines. A hybrid of small growth, but an abundant bloomer.

P. *denticulatum*, toothed-leaved Stork's-bill, S. G. 109, a shrubby much-branched stem, with palmately bipinnatifid leaves, and small purplish flowers in no great abundance. A hardy free-growing plant, but not ornamental.

P. *Smithii*, Smith's Stork's-bill, S. G. 110, a shrubby branching stem, with cordate, concave, five-lobed leaves, and dark red and light red flowers of a fine showy appearance. It is a free-growing hybrid, and an abundant bloomer.

P. *rigescens*, stiff-leaved Stork's-bill, S. G. 112, a shrubby branching stem, with roundly cordate concave leaves, and pale red flowers finely varied by dark lines. A curious hybrid, of easy culture.

PELARGONIUM *hirtum*, hairy Stork's-bill, S. G. 113, a shrubby succulent stem, with decompound leaves, and numerous linear bluntish leaflets, the flowers small and of a pale and deep red. A curious succulent species, requiring care as to water.

P. *tripartitum*, brittle-stalked Stork's-bill, S. G. 15, a shrubby branching stem, with glaucescent ternate fleshy leaves and long-petalled red and white flowers; the plant rather tender.

P. *platypetalon*, broad-petaled Stork's-bill or Oldenburgh Geranium, S. G. 116, a shrubby branching stem, with cordate, slightly lobed, unequally toothed leaves, and fine large white flowers beautifully streaked with brownish red or purple : the plant hardy and of easy culture.

P. *Scarboroviæ*, Countess of Scarborough's Stork's-bill, or Lady Scarborough's Geranium, S. G. 117, a shrubby branching stem, with small subtrifid or deeply three-lobed leaves, and fine dark and light red and white flowers. The plant flowers early in spring and late in autumn, and has a fine lemon-like scent.

P. *viscosissimum*, viscous Stork's-bill, S. G. 118, a shrubby erect branching stem, with palmately five- or seven-lobed clammy leaves, and heads of white flowers marked with light red. It is a hardy plant and an original species.

P. *formosum*, variegated-flowered Stork's-bill, S. G. 120, a shrubby branching stem, with roundly cordate, undulate, concave leaves, and a large head of dark and pale red flowers. A hardy free-flowering hybrid, in bloom from spring to autumn.

P. *amœnum*, delightful Stork's-bill, S. G. 121, a tuberous root and subcaulescent stem, with pinnate radiated leaves, and fine red flowers marked with dark lines. It is a beautiful little plant, of hybrid origin, and requiring the same treatment as the other tuberous kinds.

PELARGONIUM *Comptoniæ*, the Marchioness of North-ampton's Stork's-bill, S. G. 122, a shrubby much-branched stem, with flat, wedge-shaped, cordate leaves, and red flowers marked with darker lines and blotches. A slender growing plant, but an abundant bloomer early in spring.

P. *chrysanthemifolium*, chrysanthemum-leaved Stork's-bill, S. G. 124, a shrubby much-branched stem, with cordate, pinnatifid, or deeply seven-lobed leaves, and deep scarlet flowers marked with darker scarlet and black. A very handsome hybrid production, excelling most in brilliancy of colours. It is a free grower, and easily managed.

P. *patens*, spreading-flowered Stork's-bill, S. G. 125, a large tuberous root branching out into smaller tubers; a short, succulent, rigid, erect stem, crowned with ter-nate radiate leaves, and scarlet and white flowers. It is a curious and beautiful hybrid, easily managed.

P. *Barnardinum*, Mr. Barnard's Stork's-bill, S. G. 127, a shrubby stem with few branches, and three-part-ed, acute, smooth, a little glaucous leaves, and scarlet-veined flowers. A curious hybrid which thrives well with common treatment.

P. *candidum*, fair-flowered Stork's-bill, S. G. 128, a shrubby branching stem, with cordate three-lobed leaves, and large white flowers marked with red. A pretty plant, and an abundant bloomer.

P. *floccosum*, nappy Stork's-bill, the Jenkinson's Re-becca of gardeners, S. G. 129, a shrubby erect stem, not much branched, the leaves cordate, three-parted, and undulate, and the flower of a deep scarlet marked with black. The brilliancy of the flowers is unequalled in the genus; it requires care as to the dispensation of water, especially when not in flower.

PELARGONIUM *Watsoni*, Watson's Stork's-bill, S. G. 130, a shrubby not much branched stem, with roundly cordate five-lobed leaves, and large red flowers marked with dark or brownish red. A pretty plant, and an abundant bloomer till late in autumn.

P. *Youngii*, Young's Stork's-bill, S. G. 131, a shrubby much-branched stem, with flat widely cordate slightly three-lobed leaves, and large showy white and dark red flowers. A desirable plant, from its abundance of bloom and hardy nature.

P. *hoareæflorum*, Hoarea-flowered Stork's-bill, S. G. 133, a tuberous root, and subcaulescent succulent leafy stem; the leaves radiate, pinnate, and canescent, and the flowers deep red marked with black. It is a curious and handsome hybrid, requiring the usual culture given to the tuberous species of this genus.

P. *pulcherrimum*, beautiful Stork's-bill, S. G. 134, a shrubby erect much-branched stem, with kidney-shaped truncate leaves, and fine deep red and black flowers. It is a dwarf bushy plant, and flowers late in autumn.

P. *spectabile*, showy Stork's-bill, S. G. 136, a shrubby erect branching stem, with cordate undulate deeply-toothed leaves, and bright red flowers marked with darker red. There are several varieties of this species, all splendid plants of easy culture.

P. *Beaufortianum*, Duchess of Beaufort's Stork's-bill, S. G. 138, a shrubby erect branching stem, with rigid smoothish truncate five-lobed leaves, and large red and pale red flowers marked with black: the plant hardy and a free flowerer.

PELARGONIUM *principissæ*, Princess Charlotte's Stork's-bill, S. G. 139, a shrubby erect branching stem, with cordately reniform slightly lobed leaves, and dark red flowers marked with black blotches and lines. A beautiful plant, but rather tender; suffering however more from wet than cold.

P. *concolor*, self-coloured Stork's-bill, S. G. 140, a shrubby branching stem, with cordate five-lobed undulate leaves, and deep red flowers. There is a variety with larger flowers: both are handsome and abundant flowering plants well adapted for training to a trellis, as their shoots grow to a great length if allowed.

P. *eriophyllum*, woolly round-leaved Stork's-bill, S.G. 141, a shrubby much-branched stem with spreading branches, and flat very soft roundly cordate leaves, and pale red and white flowers: in spring the flowers are small, but they increase in size in course of the summer.

P. *Newshamianum*, Miss Newsham's Stork's-bill, S.G. 144, a shrubby branching stem, with cordate three-lobed deeply toothed leaves, and white flowers marked with red and dark brown. A pretty little plant of hybrid origin and easy culture.

P. *multiradiatum*, many-rayed Stork's-bill, S. G. 145, subcaulescent, with a tuberous large brown scaly root, short stem, and large variable deeply cut leaves; the flowers in a radiated umbel, dark brown or black, and edged with a greenish yellow. It grows freely in turfy loam, peat, and sand, and increases freely from tubers of the roots.

P. *Brownii*, Brown's Stork's-bill, Brown's Miss Rosa, S. G. 146, a shrubby short stiff branching stem, with rigid acute concave nerved leaves, and large pale red flowers marked with dark brown or purple. A hybrid as hardy as any of the genus.

PELARGONIUM *Potteri*, Potter's Stork's-bill, S.G. 147, a shrubby rather succulent branching stem, with leaves slightly cordate, deeply three-cleft, and many-nerved; the flowers of a deep scarlet finely marked with black. It is a hybrid, apt to lose its leaves in winter if not kept in a warm situation and dry: " but in summer," Mr. Sweet observes, " it grows very luxuriantly, and its flowers are much finer when growing out in the open air: they are produced till late in autumn." *Geran.* 147.

P. *pallens*, cream-coloured Stork's-bill, S. G. 148, a suffruticose short succulent stem, with deeply three-parted hairy leaves and pale yellow flowers. A rare and curious species, but not of difficult cultivation.

P. *verbenæfolium*, vervain-leaved Stork's-bill, S. G. 149, a shrubby branching stem, with ternate rough hairy leaves, and small narrow-petalled pale red flowers marked with darker red. A curious hybrid and abundant bloomer, introduced in collections for its singularity.

P. *Robinsoni*, Robinson's Stork's-bill, S. G. 150, a shrubby erect stem, with very large leaves four or five inches long and wide, cordate, acute and undulate; the flowers large, pale red marked with darker red blotches and veins. It is a hybrid of robust growth and prolific in bloom.

P. *atropurpureum*, dark purple Stork's-bill, S. G. 152, a shrubby branching stem with cordate, acute, three-lobed undulate leaves, and purple flowers marked and veined with darker purple.

P. *pyrithriifolium*, feverfew-leaved Stork's-bill, S. G. 153, a shrubby much-branched stem, with palmately bipinnatifid cordate leaves, and deep red or scarlet orange flowers marked with black. A curious hybrid flowering finely in autumn.

P. *Jenkinsoni*, Mr. Jenkinson's Stork's-bill, or John

(F) 2

Bull Geranium, S. G. 154, a shrubby branching stem, with rigid roundly cordate leaves and pale red white and very dark purplish red flowers. An elegant free-growing, abundant-flowering hybrid.

PELARGONIUM *lepidum*, pretty Stork's-bill, S. G. 156, a shrubby, much-branched stem, with cordate five-lobed undulate leaves, and pale red flowers veined with darker red. It is an elegant little hybrid, and continues in bloom till late in autumn.

P. *Tibbitsianum*, Mr. Tibbit's Stork's-bill, S. G. 158, a shrubby branching stem, with cordate slightly three-lobed undulate leaves, and large red and dark red flowers. A magnificent hybrid, and one of the largest-flowering kinds of the bright reds which has yet appeared: the plant of tolerably strong growth.

P. *æmulum*, rival Stork's bill, S. G. 160, a shrubby much-branched stem, with leaves between cordate and reniform, about as wide as long, and large light and dark red flowers veined and shaded with darker red or brown. A hardy plant, free grower, and abundant flowerer.

P. *quinquevulnerum*, dark-flowered Stork's-bill, S. G. 161, a suffruticose flexuose stem, with ternate bipinnatifid leaves, and dark brown or purple flowers edged with light red. A handsome plant, requiring care as to watering.

P. *crenulatum*, crenulate-leaved Stork's-bill, S. G. 162, a shrubby erect stem not much branched; leaves large, reniform, shallowly notched with rounded teeth; the flowers large, purplish red shaded and veined with darker colours.

P. *Murrayanum*, Lord James Murray's Stork's-bill, S. G. 164, a tall erect branching stem, with large cordate broad leaves, and fine red flowers marked with darker red or brown. It is a strong handsome plant, an abundant bloomer, and continues in flower the greater part of the year.

PELARGONIUM *ternatum*, ternate-leaved Stork's-bill, S. G. 165, B. M. 413, a shrubby much-branched stem, with distichous ternate concave rough rigid leaves, and pale red veined flowers: introduced from the Cape of Good Hope in 1789.

P. *venustum*, comely Stork's-bill, S. G. 167, a shrubby erect stem, not much branched; the leaves kidney-shaped, slightly lobed and toothed; the flowers large, white, and pale blush with dark blotches on the two upper petals. A beautiful compact bush, producing great plenty of flowers.

P. *tomentosum*, Pennyroyal Stork's-bill, or pepper-mint-scented Geranium, S. G. 168, B. M. 518, a shrubby thick succulent stem much branched, with cordate five-lobed subhastate leaves, and small white and purple flowers. An old inhabitant of the green-house, having been introduced from the Cape of Good Hope in 1790.

P. *asperifolium*, rough-leaved Stork's-bill, S. G. 169, a shrubby branching stem, with cordate deeply-lobed roughish leaves, and deep red flowers lined and shaded with darker red. A hardy ornamental plant, and a good flowerer.

P. *cruentum*, blood-red Stork's-bill, S. G. 170, a tuberous scaly root, with pinnatifid or deeply laciniate scaly leaves and deep dark red flowers. An elegant little hybrid which flowers all the summer.

P. *fragrans*, nutmeg-scented Stork's-bill, S. G. 172, a shrubby erect stem much branched, with roundly cordate three-lobed leaves, and small pale flowers tinged with blue. It continues in bloom the greater part of the year.

P. *coruscans*, glittering Stork's-bill, S. G. 173, a

shrubby much-branched stem, with slightly cordate ovate leaves, and fine deep red flowers veined and shaded with blackish red or brown. A handsome hybrid which continues in bloom the greater part of the year.

PELARGONIUM *Wellsianum*, Mr. Wells's Stork's-bill, S. G. 175, a shrubby erect branched stem, with flat cordate hairy leaves, and bright orange scarlet flowers edged with purple. One of the most splendid hybrids that has yet appeared, its brilliancy of colouring surpassing all art to imitate.

P. *ramulosum*, small-branched Stork's-bill, S. G. 177, a shrubby erect much-branched stem, with small cordate or cuneate leaves, and flowers white with large blotches of deep red and black. It is a bushy plant, a hybrid of the easiest culture.

P. *Fairlieæ*, Mrs. Fairlie's Stork's-bill, S. G. 178, a shrubby erect much-branched stem, with small reniform deeply three-lobed leaves, and reddish purple flowers with dark pink blotches. A hybrid from Paris, a late flowerer, but of easy culture.

P. *glauciifolium*, horn poppy-leaved Stork's-bill, S. G. 179, a large tuberous root, suffruticose stem, and ternate pinnatifid lobate or sinuate leaves, with dark velvet coloured flowers edged with a greenish yellow. A handsome and very curious hybrid; the flowers exquisitely fragrant, their fragrance beginning about five or six o'clock in the evening, and continuing till about nine the next morning.

P. *flexuosum*, zigzag-stalked Stork's-bill, S. G. 180, a shrubby succulent brittle flexuose stem, with cordately ovate deeply incised nerved leaves, and deep scarlet and black flowers. A beautiful hybrid, newly originated at Mr. Colville's nursery, and which Mr. Sweet says will probably flower all the winter.

PELARGONIUM *Broughtoniæ*, Lady Broughton's Stork's-bill, S. G. 181, a shrubby branching stem, with cordate five-lobed undulate leaves, and brilliant flame-coloured flowers with dark shades and veins. A handsome-growing plant, blooming all the summer and till late in autumn.

P. *lyrianthinum*, royal purple Stork's-bill, or More's Princess of Denmark Geranium, S. G. 183, a shrubby much-branched stem, with flat cordate five-lobed leaves, and large bright rosy purple flowers. A hybrid, an abundant flowerer and very hardy.

P. *acutilobum*, sharp-lobed Stork's-bill, S. G. 184, a shrubby branching stem, with flat cuneate or lanceolate leaves, and fine white flowers blotched with red and purple. The plant hardy and of free growth.

P. *interlentum*, interwoven Stork's-bill, S. G. 185, a tuberous root, the stem very short, with leaves variable, ovate, obtuse, jagged or ternate, and bright scarlet flowers veined with black. An elegant hybrid, flowering the greater part of the summer.

P. *paucidentatum*, distant-toothed Stork's-bill, S. G. 186, a shrubby flexuose branching stem, with broad cordate slightly three-lobed leaves, and bright lilac flowers tinged with rose in the centre, on which is a black velvety mark. A hardy plant, producing abundance of bloom.

P. *erectum*, upright Stork's-bill, S. G. 187, an erect suffruticose stem, with cordate jagged lobate leaves, and rosy-coloured and white flowers. The plant succulent, and requiring but little water.

P. *cosmianum*, perfumed Stork's-bill, S. G. 189, a shrubby branching stem, with small distant two-ranked deeply lobed leaves, and white flowers with dark spots on a red ground. A handsome plant with an agreeable fragrance.

Pelargonium *selectum*, choice Stork's-bill, S. G. 190, a tuberous root, and suffruticose stem, with cordate deeply five-lobed unequally toothed leaves, and rosy purple flowers with dark velvet spots nearly the size of the petals. The plant rare, but not of difficult culture.

P. *Beadoniæ*, Mrs. Beadon's Stork's-bill, S. G. 191, a shrubby upright much-branched stem, with cuneate deeply three-lobed crenate leaves, and bright reddish blue flowers with black veins. A hardy hybrid of easy culture.

P. *crassicaule*, thick-stalked Stork's-bill, S. G. 192, B. M. 477, a shrubby succulent branching stem, with reniform acuminated plicate leaves, and white flowers marked with bright purple. A fine species, introduced from the south-west coast of Africa: there are several varieties, all of them scarce, as the plants make but little wood to spare for cuttings.

P. *inscriptum*, marked-petalled Stork's-bill, S. G. 193, a shrubby branching stem, with cordate lobed toothed and very variable leaves; the flowers blush-coloured with bright red spots and finely reticulated lines which spread over the petals. A bushy plant, well adapted for a small green-house.

P. *affluens*, numerous-flowered Stork's-bill, S. G. 194, a shrubby much-branched stem, with small cordate deeply three-lobed leaves, and bright lilac flowers tinged with red and a purple mark in the centre. A bushy plant, and an abundant bloomer all the summer.

P. *dependens*, pendent-petalled Stork's-bill, S. G. 195, a shrubby branching stem, with small deeply three-lobed truncate leaves, and clear white flowers with large dark purple blotches edged with red. A handsome-flowering free-growing plant.

P. *aurantiacum*, orange-coloured Stork's-bill, S. G. 198,

a shrubby flexuose stem, with cordate lobed unequally toothed leaves, and reddish orange flowers marked with purple stripes. There are two varieties, all abundant flowerers in autumn.

PELARGONIUM *calocephalon*, pretty-headed Stork's-bill, or Tull's Imperial Geranium, S. G. 201, a shrubby branching stem, with flat cordate acute deeply five-lobed leaves, and pale blush flowers marked with red spots and numerous branching lines. A fine strong growing plant, and an abundant bloomer from May to November.

P. *incanescens*, whitish-leaved Stork's-bill, S. G. 203, a shrubby erect branched stem, with cordate deeply lobed toothed leaves, and pale lilac flowers marked with dark purple spots and numerous purple lines. A fine strong-growing hybrid.

P. *modestum*, modest Stork's-bill, S.G. 204, a shrubby erect much-branched stem, with small cordate three-lobed undulate leaves, and pale blush-coloured flowers marked with dark lines and red blotches. The plant grows to the height of three feet, and flowers abundantly all the summer.

The culture of this genus has been already given, Part I. p. 76.

MELIACEÆ.

EKEBERGIA *capensis*, Cape Ekebergia, a shrub, a native of the Cape of Good Hope, introduced in 1789. It grows in loam and peat, and may be increased by cuttings of the young wood in sand under a bell-glass.

MELIA *Azedarach*, common Bead-tree, B. M. 1066, a deciduous tree, introduced from the South of Europe in 1656, and flowering from June to August. It grows in loam and peat, and ripened cuttings root in sand under a hand-glass, their leaves not being taken off.

AITONIA *capensis*, Cape Aitonia, B. M. 173, B. C. 682, a shrub from the Cape of Good Hope, introduced in 1774, and flowering from April to September.

AURANTEÆ.

MURRAYA *exotica*, ash-leaved Murraya, B. R. 434, a tree from India, introduced in 1771, and flowering in August and September.

CAMELLIA *Bohea*
———— *viridis*
———— *Sasanqua*, small white flower
—————————*rosea*, fine rose-coloured Palmer's Camellia
———— *japonica rubra*, single red
———————— *alba*, single white
———————— *semiduplex*, semi-double red
———————— *rubra plena*, old double red
———————————————— young double red
———————————— superb double red
———— *carneo-plena*, Middlemist's blush
———— *myrtifolia*, myrtle-leaved red-flowering
———— *involuta*, Lady Long's, red flower
———— *atro-rubens*, Loddiges' red
———— *anemoneflora*, Waratah red
———————————— *variegata*, carnation Waratah
———————————— *carnea*, blush Waratah
———— *pæoniflora*, pæony-flowered
———— *variegata*, double striped
———— *pomponia*, Kew blush
———— *flavescens*, Lady Hume's blush
———— *luteo-alba*, Bassington's yellowish white
———— *alba pleno*, double white
———— *longifolia*, long-leaved
———— *rubricaulis*, Lady Admiral Campbell's

Camellia *Wellbankiana,* Wellbank's Camellia, double white
———— *conchiflora*
———— *oleifera,* said to be a distinct species from *japonica*
———— *hexangularis,* double red
———— *axillaris.*

There is a yearly increase of new sorts of *Camellia* from hybrid seedlings, all of them of great beauty both in flower and foliage. The culture of the genus has been already given, Part I. p. 39.

Citrus *medica,* the Lemon-tree, a tree introduced from Asia in 1648, and flowering from May to July.

C. *acida,* the Lime-tree, a tree introduced from Asia in 1648, and flowering from May to July.

C. *Aurantium,* the common Orange-tree, a fruit-tree introduced from India in 1595, and flowering from May to July.

C. *sinensis,* myrtle-leaved Orange-tree, a tree from China.

C. *buxifolia,* box-leaved Orange-tree, a tree from China.

C. *nobilis,* Mandarin Orange-tree, A. R. 608, a tree introduced from China in 1805, and flowering from May to July.

C. *nobilis minor,* smaller Mandarin Orange-tree, B. R. 211.

C. *decumana,* Shaddock-tree, a tree introduced from India in 1724, and flowering from May to July.

The culture of the *Citrus* tribe has been already given, Part I. p. 81.

RUTACEÆ.

Zygophyllum *cordifolium,* heart-leaved Bean-caper, a low shrub introduced from the Cape of Good Hope in 1774, and flowering in October.

ZYGOPHYLLUM *fœtidum*, fœtid Bean-caper, B. M. 372, a shrub introduced from the Cape of Good Hope in 1790, and flowering from June to August.

Z. *maculatum*, spotted-flowered Bean-caper, a shrub introduced from China in 1782, and flowering in October and November.

Z. *Morgsana*, four-leaved Bean-caper, a shrub introduced from the Cape of Good Hope in 1732, and flowering from May to September.

Z. *sessilifolium*, sessile-leaved Bean-caper, a shrub introduced from the Cape of Good Hope in 1713, and flowering in July and August.

These are ugly plants, easily grown in loam and peat, and cuttings root under a hand-glass.

FAGONIA *cretica*, Cretan Fagonia, B. M. 241, an annual from Candia introduced in 1739, and flowering from June to August.

F. *arabica*, Arabian Fagonia, a biennial introduced from Arabia in 1759, and flowering from June to August.

These plants, which are void of beauty, grow in any light rich soil: they are best raised in a hot-bed, and then transplanted and placed among the green-house plants.

RUTA *chalepensis*, broad-leaved African Rue, a low evergreen shrub introduced from Africa in 1722, and flowering from June to September.

R. *angustifolia*, narrow-leaved African Rue, a very low evergreen shrub introduced from Africa in 1722, and flowering from June to September.

R. *pinnata*, winged-leaved Rue, a low evergreen shrub from the Canaries in 1780, and flowering from March to August.

Ugly fœtid plants which grow in any dry soil, and increase with care by cuttings or seeds.

MELIANTHUS *major*, great Honey-flower, B. R. 45,

a suffruticose or spongy-wooded stem, with large glaucous pinnatifid leaves and white mellifluous flowers: introduced from the Cape of Good Hope in 1688, and flowering from May to July.

MELIANTHUS *minor*, small Honey-flower, B. M. 301, a plant similar to the other, but with smaller leaves. It was introduced from the Cape of Good Hope in 1696, and flowers in August.

These plants are showy from their large glaucous leaves: they grow in any light rich soil, and are readily increased by cuttings or suckers.

DIOSMEÆ.

ADENANDRA *uniflora*, one-flowered Adenandra, B. M. 273, a shrub introduced from the Cape of Good Hope in 1775, and flowering from April to June.

A. *umbellata*, umbel-flowered Adenandra, B. M. 1271, a shrub from the Cape in 1789, flowering from April to June.

A. *fragrans*, red-flowered Adenandra, B. M. 1519, a shrub introduced from China in 1812, flowering from May to July.

A. *tetragona*, tetragonal Adenandra, a shrub introduced from the Cape of Good Hope in 1789, and flowering in July and August.

A. *amœna*, charming Adenandra, B. R. 553, B. C. 161, a shrub from the Cape in 1790, flowering in July and August.

These are elegant evergreen plants which grow in sandy peat, or peat and loam with sand; and they are increased by cuttings of the young tender tops, like the Heaths, Part I. p. 57.

BAROSMA *serratifolia*, saw-leaved Barosma, B. M.

456, B. C. 373, a shrub from the Cape of Good Hope in 1789, flowering from March to June.

BAROSMA *latifolia*, broad-leaved Barosma, A. R. 33, B. C. 290, a shrub from the Cape in 1739, flowering in July and August.

B. *crenata*, crenated Barosma, B. C. 404, a shrub from the Cape in 1774, and flowering from January to March.

B. *ovata*, oval-leaved Barosma, B. M. 1616, a shrub from the Cape in 1790, flowering from February to September.

Beautiful evergreen plants, some of them odoriferous, and others showy when in bloom. Their culture is the same as for *Adenandra*.

DIOSMA *oppositifolia*, opposite-leaved Diosma, an evergreen shrub from the Cape in 1752, and flowering from March to June.

D. *linearis*, linear-leaved Diosma, an evergreen shrub from the Cape in 1800, and flowering from March to June.

D. *alba*, white-flowered Diosma, also from the Cape, and flowering from March to June.

D. *hirsuta*, hairy-leaved Diosma, B. R. 369.

D. *rubra*, red-calyxed Diosma, B. R. 563.

D. *pectinata*, pectinated Diosma.

D. *ericoides*, heath-leaved Diosma.

D. *cupressina*, cypress-leaved Diosma, B. C. 303.

D. *tenuifolia*, slender-leaved Diosma.

D. *succulenta*, succulent-leaved Diosma.

The last seven species are all from the Cape, and flower from March or April to May or June. They are all much of the nature of Heaths, thrive best in peat soil, and are increased by cuttings of the tender tops planted in sand under a bell-glass.

AGATHOSMA *hispida,* rough-leaved Agathosma, a nar-row-leaved evergreen shrub from the Cape in 1786, and flowering from June to August.

A. *villosa,* shaggy Agathosma, a similar shrub from the Cape in 1786, and flowering at the same time.

A. *pulchella,* blunt-leaved Agathosma, B. M. 1357, a shrub from the Cape in 1787, and flowering from February to September.

A. *ciliata,* ciliated Agathosma, B. R. 366.

A. *capitata,* pale purple Agathosma.

A. *imbricata,* imbricated Agathosma.

A. *acuminata,* acuminated Agathosma, B. C. 493.

A. *marginata,* marginate Agathosma.

A. *cerefolia,* chervil-scented Agathosma.

A. *pubescens,* pubescent Agathosma.

The last seven species are all from the Cape, and all flower in May or June to August. The culture of the whole genus is the same as that for *Diosma.*

CORRÆA *alba,* white-flowered Corræa, A. R. 18, B. R. 515, B. C. 152, a shrub from New South Wales in 1793, and flowering from April to June.

C. *speciosa,* red-flowered Corræa, B. R. 26, B. C. 112, B. M. 1746, an evergreen, flowering at the same period as C. *alba.*

C. *virens,* green-flowered Corræa, B. R. 3, B. C. 336, an evergreen shrub from New South Wales in 1800, flowering from November to May.

These are handsome and very hardy plants; so hardy that they bear the open air all the year in Jersey and Guernsey: they grow in loamy soil, are increased readily by seeds or cuttings; and C. *speciosa,* which is a very fine plant, is sometimes grafted on C. *alba.*

EMPLEURUM *serrulatum,* Cape Empleurum, a shrub from the Cape in 1774, and flowering in June and July.

It is a plant of little beauty, but of easy culture in sandy peat; and cuttings of the young wood root readily in the same soil.

TEREBINTACEÆ.

CNEORUM *tricoccum*, smooth Widow-wail, a low shrub from the South of Europe in 1793, producing small yellow flowers from April to September. It grows in loamy soil, and is increased by cuttings of the young wood in the same soil or by seeds. It is a dull ugly plant.

FAGARA *Piperita*, ash-leaved Fagara, a tree from Japan in 1773, flowering in September. The plant is without beauty, but grows freely in sandy loam, and is increased by young cuttings in the same soil.

RHUS *succedanum*, red Lac Sumach, a shrub from China in 1768, and flowering in June.

R. *semialatum*, service-leaved Sumach, a shrub from Macao in 1780.

R. *pendulinum*, pendulous Sumach.

R. *dentatum*, rough-stalked Sumach.

R. *cuneifolium*, wedge-leaved Sumach.

R. *incisum*, cut-leaved Sumach.

R. *tomentosum*, woolly-leaved Sumach.

R. *villosum*, hairy Sumach.

R. *pubescens*, pubescent Sumach.

R. *viminale*, willow-leaved Sumach.

R. *angustifolium*, narrow-leaved Sumach.

R. *rosmarinifolium*, rosemary-leaved Sumach.

R. *pentaphyllum*, various-leaved Sumach.

R. *lævigatum*, polished-leaved Sumach.

R. *lucidum*, shining-leaved Sumach.

R. *lucidum minus*, small shining-leaved Sumach.

All these plants are natives of the Cape of Good Hope, excepting R. *pentaphyllum*, which is from Barbary: few of them have yet flowered in this country; but those which have, bloom in July and August. They are chiefly weak deciduous plants, with compound leaves of no beauty whatever. They grow in sandy loam, and are increased by cuttings of the ripe wood.

DODONÆA *triquetra*, three-sided Dodonæa, A. R. 230, a shrub from New South Wales in 1790, flowering from June to August.

D. *angustifolia*, narrow-leaved Dodonæa, a shrub from the Cape in 1758, flowering from May to August.

These plants, which are of little beauty, grow in loam and peat, and are increased by young cuttings in sand under a bell.

PISTACIA *atlantica*, Atlantic Pistachia-tree, a small tree from Barbary in 1790.

P. *Lentiscus*, Mastic-tree, B. M. 1967, a small tree from the South of Europe in 1664, flowering in May.

These plants have fine compound leaves and are evergreens; but they flower sparingly and are not ornamental: they grow in sandy loam, and are increased by young cuttings in sand under a bell-glass.

XANTHOXYLUM *Clava Herculis*, lentiscus-leavedToothach-tree, a low tree from the West Indies in 1739, flowering in April and May. It grows in sandy loam, and is increased by cuttings of the young wood in the same soil and covered with a bell-glass.

SCHINUS *mollis*, Peruvian Schinus, a shrub from Peru in 1597, flowering in July and August.

S. *dentata*, toothed-leaved Schinus, A. R. 620, a shrub from Owhyhee in 1795, flowering from May to July.

These plants, of no beauty, grow in loam and peat, and are increased by cuttings in sand under a bell-glass.

RHAMNEÆ.

ILEX *crocea,* African Holly, a shrub from the Cape of Good Hope in 1794, flowering in May and June.

I. *Perado,* thick-leaved Holly, a shrub from Madeira in 1760, flowering in April and May.

I. *Cassine, Dahoon,* and *angustifolia,* are frame shrubs from Carolina, and I. *vomitoria* is a frame shrub from Florida; all which flower in April and May.

All the above species grow in loamy soil, and are increased by ripened cuttings in sandy loam under a hand-glass.

ELÆODENDRUM *australe,* thick-leaved Olive-wood, a small tree from New South Wales in 1796, flowering from June to August. It grows in loam and peat, and ripened cuttings root in sand under a hand-glass.

RHAMNUS *lycioides,* boxhorn-like Buckthorn, a shrub from Spain in 1782, flowering from September to December.

R. *crenulatus, Theezans, tetragonus, latifolius, glandulosus,* and *prinoides,* are species which are inhabitants of the green-house, but without a single recommendation. They grow in loam and peat, and ripened cuttings root freely in sand under a frame or glass.

ZIZYPHUS *lineatus,* lined Zizyphus, a shrub from China in 1804, which has not yet flowered.

Z. *Lotus,* a small tree from Barbary, supposed by some to be the Lote-tree of Pliny.

Z. *vulgaris,* common Zizyphus, a small tree from the South of Europe, which flowers in August and September.

These plants are of no beauty, and may be treated like *Rhamnus.*

CELASTRUS *lucidus,* shining Stuff-tree a low tree from

the Cape of Good Hope, flowering from April to September.

CELASTRUS *Pyracanthus*, pyracantha-leaved Staff-tree, a low tree also from the Cape, flowering in May and June.

C. *octagonus*, *cassinoides*, *cernuus*, *tetragonus*, and *buxifolius*, are also green-house species of this genus, but all of them of little show or beauty. They are propagated like *Ilex*.

EUONYMUS *japonicus*, Japan Spindle-tree, a low tree from Japan, flowering from June to August. It grows in sandy loam, and is increased by cuttings of the young wood in the same soil.

HOVENIA *dulcis*, sweet Hovenia, a shrub from Japan, introduced in 1812. It may be treated like *Euonymus*.

CEANOTHUS *microphyllus*, small-leaved Ceanothus, a shrub from North America, flowering from June to August.

C. *africanus* and *globulosus* are, like the first species, obscure flowering plants of no beauty, which may be treated like *Ilex*.

POMADERRIS *apetala*, petalless Pomaderris, a shrub from New Holland, flowering in May and June.

P. *elliptica*, oval-leaved Pomaderris, B. M. 1510, and P. *lanigera*, woolly-leaved Pomaderris, B. M. 1823, B. C. 357, are also New Holland shrubs, flowering in May and June. They are of little beauty, grow in loam and leaf-mould, and root in sand under a glass.

PHYLICA *ericoides*, heath-leaved Phylica, B. M. 224, a small heath-looking shrub from the Cape, flowering from September to April in a warm green-house.

P. *parviflora*.	P. *pubescens*, B. C. 695.
lanceolata.	eriophoros.
capitata, B. R. 711.	rosmarinifolia.

(84)

PHYLICA *axillaris.*	PHYLICA *myrtifolia.*
plumosa, B. C. 253.	*callosa.*
villosa.	*paniculata.*
stipularis.	*imbricata.*
cordata.	*cylindrica.*
buxifolia.	*racemosa.*
spicata, B. C. 323.	*pinifolia.*

These are all low heath-looking evergreens, generally with dull white flowers, which appear from April or May to September. They grow in sandy peat, and are propagated in the same manner as heaths, but with much less trouble.

BRUNIA *nodiflora,* imbricated Brunia, a low shrub, with heath- or fir-like leaves, flowering in July and August.

B. *paleacea.*	B. *alopecuroides.*
lanuginosa, B. C. 572.	*abrotanoides,* B. C. 355.
verticillata.	*superba.*
deusta.	*fragarioides.*
microphylla.	*ciliata.*
laxa.	

These are all evergreens from the Cape, handsome bushy plants, though without showy flowers : they all grow in peat soil which must not be over-watered ; and they are increased in the same manner as Heaths.

STAAVIA *radiata* and *glutinosa* are Cape shrubs resembling *Brunia,* and requiring similar treatment.

PLECTRONIA *corymbosa,* corymbed Plectronia, is a shrub from the Cape in 1816, which may be treated like *Brunia.*

CASSINE *capensis,* Cape Phillyrea,

C. *Colpoon,* the Colpoon tree,

C. *lævigata,* smooth Cassine, are shrubs from the Cape, which seldom flower in this country.

CASSINE *Maurocenia*, the Hottentot Cherry, is also from the Cape, and flowers in July and August.

All the species grow in loamy soil with a little peat- or leaf-mould, and cuttings root in sand under a hand-glass.

PRINOS *lucidus*, shining Winter-berry, a frame shrub, flowering in June and July, which may be treated like *Cassine*.

LEGUMINOSÆ.

EDWARDSIA *grandiflora*, large-flowered Edwardsia, B. M. 167.

E. *chrysophylla*, golden-leaved Edwardsia, B. C. 738, B. M. 1442.

E. *microphylla*, small-leaved Edwardsia.

These are pretty little frame shrubs, which grow readily in sandy loam, flower freely and ripen seeds, from which, or from young cuttings under a bell-glass, they are usually propagated.

ANAGYRIS *fœtida*, B. C. 740, and *latifolia*, Bean Trefoils, the first a frame shrub from Spain, and the other a green-house shrub from Teneriffe, flowering in April and May. They grow in loam and peat, and are increased by cuttings in the same soil.

VIRGILIA *aurea*, great-flowered Virgilia, an Abyssinian shrub which flowers in July.

V. *intrusa* and V. *capensis*, B. M. 1590, Cape shrubs which flower from May to August.

These are rather pretty plants, which grow in loam and peat, and young cuttings root in covered pots of sand.

CYCLOPIA *genistoides*, genista-leaved Cyclopia, B. M. 1259, a Cape shrub flowering in July and August; a handsome plant, which grows in sandy loam and peat, and very young cuttings will root in sand under a bell, care being taken to prevent their damping off.

BAPTISIA *perfoliatu,* perfoliate Baptisia, a perennial from Carolina, flowering in August. It is a plant of no beauty, growing in peat loam, and increased by cuttings or dividing at the root.

PODALYRIA *myrtifolia,* myrtle-leaved Podalyria.

P. *sericea,* silky Podalyria, B. M. 1923.

P. *cuneifolia,* wedge-leaved Podalyria.

P. *biflora,* two-flowered Podalyria, B. M. 753.

P. *calyptrata,* one-flowered Podalyria.

P. *styracifolia,* storax-leaved Podalyria, B. M. 1580.

P. *buxifolia,* box-leaved Podalyria, B. C. 649.

P. *oleæfolia,* olive-leaved Podalyria.

P. *hirsuta,* hairy Podalyria.

P. *cordata,* heart-leaved Podalyria.

These are all pretty Cape plants, flowering from March to June: they grow in loam and peat, are increased by ripened cuttings, and sometimes perfect seeds.

CHORIZEMA *ilicifolia,* holly-leaved Chorizema.

C. *nana,* dwarf Chorizema, B. M. 1032.

C. *rhombea,* few-flowered Chorizema.

These are pretty New Holland shrubs, flowering from February to June or later. They grow in sandy loam and peat, and produce abundance of seeds.

PODOLOBIUM *trilobatum,* B. M. 1477, a shrub from New South Wales, flowering from April to July; and which may be treated like *Chorizema.*

OXYLOBIUM *arborescens,* B. R. 392, B. C. 163, and O. *ellipticum,* are Van Dieman shrubs.

O. *cordifolium,* A. R. 494, B. M. 1544, is a New Holland shrub. The three species flower from April to August; they grow in sandy loam and peat, and may be increased by young cuttings in sand under a bell, or by seeds which they sometimes ripen.

Callistachys *lanceolata*, B. R. 216.

C. *ovata*, B. M. 1925.

These are New Holland shrubs, which flower from June to August; they grow rapidly, flower freely in loam and peat soil, and are readily increased by cuttings or seeds.

Brachysema *latifolium*, B. M. 2008, B. C. 411, B. R. 118, a New Holland shrub, which grows freely in common garden soil, and flowers from April to July.

Gompholobium *latifolium*, broad-leaved Gompholobium.

G. *grandiflorum*, large-flowered Gompholobium, B.R. 484.

G. *marginatum*, small-flowered Gompholobium.

G. *polymorphum*, variable Gompholobium, B.M.1533.

G. *minus*, hairy-stalked Gompholobium.

G. *tomentosum*, tomentose Gompholobium.

G. *venustum*, purple-flowered Gompholobium.

These are pretty Australasian plants, which flower from March to August: they grow in sandy loam and peat, and are increased by young cuttings in sand and covered with a bell-glass.

Burtonia *scabra*, rough-leaved Burtonia, is a handsome New Holland shrub, flowering from May to July. Sweet observes, that " it requires more than ordinary treatment to keep it in good health." He recommends equal parts of sandy loam and peat, and well drained pots. Young cuttings, he says, are not difficult to root in sand under a bell-glass. Sometimes it ripens seeds.

Jacksonia *scoparia*, B. C. 427, and *spinosa*, are pretty New Holland plants, which flower from May to August; they grow in loam and peat, and are increased by young cuttings in sand under a bell, or ripened cuttings in sand under a hand-glass.

VIMINARIA *denudata*, leafless Rushbroom, B. M.
1190, a New Holland shrub, flowering from June to
September. It grows in sandy loam and peat, and is
increased either by seeds or cuttings.

SPHÆROLOBIUM *vimineum*, B. M. 969, and *medium*,
are pretty New Holland shrubs, flowering from May to
August : they grow in sandy loam and peat, and are
increased by young cuttings in sand under a bell-glass,
or by seeds.

AOTUS *villosa*, B. M. 969, a pretty New Holland
shrub, flowering from May to August, which may be
treated like *Sphærolobium*.

DILLWYNIA *floribunda*, close-flowered Dillwynia,
B. C. 305.

D. *ericifolia*, heath-leaved Dillwynia.

D. *glaberrima*, smooth Dillwynia, B. M. 944, B. C.
582.

D. *parvifolia*, small-leaved Dillwynia, B. M. 1527.

These are handsome New Holland shrubs with heath-
like leaves, and flowering from March to June. They
grow best in very sandy loam and peat with well-drained
pots, and are propagated like *Erica*.

EUTAXIA *myrtifolia*, B. M. 1274. This is a pretty
New Holland shrub, flowering from March to June : it
grows in sandy loam and peat, and as it grows fast and
tall, should be well cut in to form it into a bushy shrub :
young cuttings root in sand under a glass.

SCLEROTHAMNUS *microphyllus*, a New Holland shrub,
flowering in May and June, which requires similar
treatment to *Eutaxia*.

GASTROLOBIUM *bilobum*, B. R. 411, B. C. 70, a pretty
New Holland shrub, flowering from March to May, and
of the same culture as *Eutaxia*.

EUCHILUS *obcordatus*, B. C. 60, B. R. 403, a pretty

little New Holland shrub, flowering from March to June: it grows in sandy loam and peat, and young cuttings root in sand under a bell.

PULTENÆA *daphnoides,* daphne-leaved Pultenæa, B. M. 1394.

P. *obcordata,* heart-leaved Pultenæa, A. R. 574.

P. *scabra,* rough-leaved Pultenæa.

P. *retusa,* blunt-leaved Pultenæa, B. R. 378.

P. *stricta,* upright Pultenæa, B. M. 1588.

P. *linophylla,* flax-leaved Pultenæa.

P. *paleacea,* chaffy Pultenæa, B. C. 291.

P. *stipularis,* scaly Pultenæa, B. M. 475.

P. *vestita,* awned Pultenæa.

P. *villosa,* villous Pultenæa, B. M. 967.

P. *flexilis,* shining-leaved Pultenæa.

Australasian shrubs, which flower from April and May to July, grow in sandy loam and peat, and are increased by cuttings under a bell-glass in sand.

DAVIESIA *acicularis,* needle-leaved Daviesia.

D. *ulexina,* furze-leaved Daviesia, A. R. 304, B.C. 44.

D. *corymbosa,* glaucous-leaved Daviesia.

D. *mimosoides,* green-leaved Daviesia, A. R. 526.

D. *latifolia,* broad-leaved Daviesia, B. M. 1757.

These are beautiful New South Wales shrubs, flowering from May to August; they grow in equal parts of sandy loam and peat, and cuttings when they begin to ripen root in sand under a bell-glass.

MIRBELIA *reticulata,* B. M. 1211, and *dilatata,* are pretty New Holland shrubs, flowering from May to August, which may be treated like *Daviesia.*

CASSIA *corymbosa,* corymbose Cassia, B. M. 633, a Buenos Ayres shrub, flowering in July.

C. *ruscifolia,* ruscus-leaved Cassia, a Madeira shrub, flowering from May to July.

CASSIA *chinensis*, Chinese Cassia, a China shrub.

C. *multiglandulosa*, glandulous Cassia, a Teneriffe shrub, flowering from June to August.

C. *capensis*, Cape Cassia, B. C. 511, a Cape shrub.

C. *nictitans*, Virginian Cassia, a North American biennial, flowering in July.

All these plants grow in loam and peat, and are increased from cuttings planted in sand under a hand-glass, or from seeds.

ERYTHRINA *herbacea*, herbaceous Coral-tree, B. M. 877, a low Carolina shrub, flowering from June to September.

E. *Caffra*, Cape Coral-tree, B. C. 736, a Cape shrub introduced in 1816.

These are handsome plants; they grow freely in a mixture of loam and peat, and cuttings somewhat ripened will root readily under a hand-glass in sand.

BORBONIA *ericifolia*, heath-leaved Borbonia.

B. *trinervia*, three-nerved Borbonia.

B. *lanceolata*, many-nerved Borbonia, B. C. 81.

B. *perfoliata*, perfoliate Borbonia.

B. *undulata*, wave-leaved Borbonia.

B. *cordata*, heart-leaved Borbonia.

B. *crenata*, notch-leaved Borbonia, B. M. 274.

These are all Cape shrubs, flowering from July to September, and growing freely in loam and peat, and increased by young cuttings in sand.

SPARTIUM *monospermum*, white single-seeded Broom, B. M. 683.

S. *sphærocarpon*, yellow single-seeded Broom.

S. *procerum*, tall Portugal Broom.

S. *congestum*, close-branched Teneriffe Broom.

S. *virgatum*, long-twigged Madeira Broom.

S. *umbellatum*, umbelled Barbary Broom.

SPARTIUM *sericeum,* silky Mogadore Broom.

S. *nubigenum,* cluster-flowered Broom.

S. *linifolium,* flax-leaved Broom, B. M. 442.

S. *ferox,* fierce Broom, B. R. 368.

S. *spinosum,* prickly Broom.

These shrubs are of little beauty : they grow in loam and peat, and young cuttings root under a bell-glass in sand. They often ripen seeds.

GENISTA *canariensis,* Canary Genista, B. R. 217.

G. *viscosa,* clammy Genista.

These shrubs flower from May to September, and are to be treated like *Spartium.*

SEBECKIA *contaminata,* B. R. 102, *sericea,* and *cytisoides,* are Cape shrubs which flower in April and May: they grow in loam and peat, and are increased by young cuttings in sand under a glass.

RAFNIA *triflora,* B. M. 482, a Cape biennial, flowering in June and July, which may be raised on a hot-bed and transplanted into any rich light soil. It is increased by cuttings or seeds.

ASPALATHUS *chenopoda,* goose-foot Aspalathus, B. M. 2225.

A. *albens,* silky Aspalathus.

A. *pedunculata,* small-leaved Aspalathus, B. M. 344.

A. *ericifolia,* heath-leaved Aspalathus.

A. *asparagoides,* asparagus-leaved Aspalathus.

A. *carnosa,* fleshy-leaved Aspalathus, B.M. 1289.

A. *crassifolia,* bristle-pointed Aspalathus, A. R. 351.

A. *ciliaris,* ciliated Aspalathus, B. M. 2233.

A. *uniflora,* single-flowered Aspalathus.

A. *subulata,* awl-leaved Aspalathus.

A. *globosa,* globular Aspalathus, A. R. C. ic.

A. *araneosa,* cobweb Aspalathus, B. M. 829.

A. *argentea,* silver-leaved Aspalathus.

ASPALATHUS *candicans*, white Aspalathus.

A. *callosa*, oval-spiked Aspalathus.

A. *mucronata*, thorny-branched Aspalathus.

These are all Cape shrubs, growing in loam and peat, and increased by young cuttings in sand. They flower for the most part from June to August.

SARCOPHYLLUM *carnosum*, a Cape shrub, rather succulent, flowering from May to August: it grows in sandy loam, and ripened cuttings root in the same soil with an addition of sand and covered with a glass.

STAURACANTHUS *aphyllus*, a frame-shrub from Portugal, flowering in June and July, and which may be treated as *Sarcophyllum*.

AMORPHA *microphylla*, *pubescens*, and *canescens*, bastard Indigos, frame-shrubs from North America, flowering from June to August. They grow in peat and loam, and are increased by cuttings or layers.

PLATYLOBIUM *formosum*, large-flowered Flat-pea, B. M. 469.

P. *parviflorum*, small-flowered Flat-pea, B. M. 1520.

P. *triangulare*, triangular-leaved Flat-pea, B.M.1508.

Australasian shrubs of some beauty, which may be treated like *Aspalathus* above.

BOSSIÆA *Scolopendrium*, plank-plant Bossiæa, B. R. 191.

B. *ruffa*, red-flowered Bossiæa.

B. *heterophylla*, various-leaved Bossiæa, . B.M. 1144.

B. *linophylla*, narrow-leaved Bossiæa.

B. *prostrata*, procumbent Bossiæa, B. M. 1493.

B. *cinerea*, downy sharp-leaved Bossiæa, B. R. 306.

B. *microphylla*, small-leaved Bossiæa, B. M. 862.

Australasian shrubs, which may be treated as *Aspalathus*: they flower chiefly in May, June and July.

SCOTTIA *dentata*, a New Holland shrub, which flowers from June to September. Culture as in *Aspalathus*.

TEMPLETONIA *retusa*, B. R. 383, B. C. 526, a New Holland shrub, which flowers from March to June, and is treated like *Scottia* or *Aspalathus*.

GOODIA *lotifolia*, B. M. 958, B. C. 696, and *pubescens*, B. M. 1310, are shrubs from Van Dieman's Land which flower from April to July, and are grown in loam and peat, and increased by young cuttings in sand under a glass.

LODDIGESIA *oxalidifolia*, B. M. 965, a Cape shrub, flowering from May to September. Culture as in *Goodia*.

WIBORGIA *sericea*, a Cape shrub, flowering in July and August. Culture as in *Goodia* above.

CROTALARIA *cordifolia, purpurea*, B. R. 128, *pulchella*, B. M. 1699, *incanescens*, and *angustifolia*, are Cape shrubs, flowering in July and August; of the usual culture in loam and peat, and propagation by young cuttings in sand.

HOVEA *linearis*, B. R. 463, *longifolia*, B. R. 614, *lanceolata*, B. M. 1624, and *ovata*, are New Holland shrubs, flowering from March till August or later.

ONONIS *cernua*, hanging-podded Rest-harrow.

O. *vaginalis*, sheathed-podded Rest-harrow.

O. *pinguis*, greasy Rest-harrow.

O. *Natrix*, yellow shrubby Rest-harrow, B. M. 329.

O. *hispanica, tridentata*, and *crispa*, are low shrubs from the Cape, Egypt, and the South of Europe: they are of no beauty, grow in loam and peat, and are increased by seeds or young cuttings in sand.

ANTHYLLIS *sericea*, winged-leaved Kidney-vetch, a biennial from Barbary, flowering in July and August.

ANTHYLLIS *Barba-Jovis*, Jupiter's-beard Kidney-vetch, B. M. 1927.

A. *cretica*, B. M. 1092, *heterophylla, cytisoides*, and *tragacanthoides*, are green-house shrubs from the South of Europe, flowering from April to July.

A. *Hermanniæ* and *Erinacea*, B. C. 318, are frame-shrubs from the South of Europe, also flowering from April to July.

All these are pretty plants, which grow in loam and peat, and seed freely, or they may be increased by young cuttings in sand under a bell-glass.

LUPINUS *villosus* and *arboreus*, B. M. 682, the first a perennial and the other a shrubby Lupin from North America: they flower in July and August, grow in sandy loam, and are increased by seeds.

DOLICHOS *hirsutus, reticulatus*, and *lignosus*, B. M. 380, are twining East Indian shrubs, which grow in sandy loam, and are increased by cuttings in the same soil, by cuttings of the root, or by seeds: they flower in July and August.

GLYCINE *angustifolia* and *bituminosa*, B. C. 261, are green-house shrubby twiners, and G. *reniformis* is a frame perennial twiner, from the Cape and America: they may be treated as *Dolichos*.

KENNEDIA *rubicunda*, B. M. 268, *coccinea, prostrata*, B. M. 210, *Comptoniana* (*Glycine Compt.* B. R. 298), and *monophylla*, B. M. 263, are New Holland shrubby twiners, flowering from March to August, and of the same culture as *Dolichos*.

VICIA *capensis* and *pellucida* are Cape perennials, which grow in peat soil, and are increased by dividing at the root or by seeds.

LIPARIA *sphærica*, B. M. 1241, B. C. 642, *capitata, tomentosa, vestita*, A. R. 382, B. M. 2223, *graminifolia*, *villosa, hirsuta,* B. C. 38, B. R. 8, and *sericea,* are Cape shrubs, which flower from June to August, grow in loam and peat, and are increased by very young cuttings in sand under a bell.

CYTISUS *tomentosus,* A. R. 237, *foliolosus,* B.M. 426, *nanus* and *proliferus,* B. R. 121, are green-house shrubs, and C. *divaricatus,* B. M. 1387, is a frame shrub, flowering from May to August, growing in loam and peat, and increased by seeds or young cuttings in sand.

SWAINSONIA *galegifolia,* B.M. 792, and *coronillifolia,* B. M. 1725, are New South Wales shrubs, flowering in July and August, growing in loam and peat, and increased by young cuttings in sand under a bell-glass.

SUTHERLANDIA *frutescens,* B. M. 181, is a Cape shrub, flowering in June and July, which may be treated as *Swainsonia.*

LESSERTIA *annua* and *diffusa* are Cape annuals, which may be sown in peat, and transplanted into pots of light rich earth for the green-house; and L. *perennans* is a Cape perennial, which flowers in August, grows in sandy loam, and is increased by cuttings or dividing at the root.

CORONILLA *juncea,* B. C. 235, *glauca,* B. M. 13, *argentea, viminalis,* and *valentina,* B.M. 185, are showy yellow-flowered shrubs from the South of Europe, flowering the greater part of the year, and growing in any sandy soil. C. *coronata,* B. C. 907, and *minima,* B. M. 2179, are perennials from the South of Europe, of easy culture in sandy loam, and like all the species increased by seeds or cuttings.

HIPPOCREPIS *balearica,* B. M. 427, a Minorca shrub, flowering in May and June, growing in sandy loam, and increased by young cuttings in the same soil.

HALLIA *flaccida* is a Cape biennial, H. *cordata* a Cape perennial, and H. *imbricáta*, B.M. 1850, B.C. 381, a Cape shrub; all flowering in August, growing in sandy loam, and increased by cuttings, seeds, or dividing at the root.

LESPEDEZA *lagopodioides*, a Chinese perennial, flowering in May and June, of common culture.

HEDYSARUM *Alhagi*, a Levant shrub, flowering from July to August.

H. *tomentosum*, a China perennial, flowering in June and July.

H. *muricatum*, a frame shrub from Patagonia, flowering in June and July.

All these plants thrive in any rich light soil, and seed abundantly, from which, or from young cuttings, they are readily propagated.

INDIGOFERA *filifolia*, B. R. 104.

I. *psoraloides*, B. M. 476.

I. *candicans*, B. M. 198.

I. *amœna, incana, sarmentosa, denudata*, B. C. 500, *microphylla, coriacea, cytisoides*, B. M. 742, *angustifolia*, B. M. 465, and *australis*, B. C. 149, are Cape, Indian, or Australasian shrubs, which grow in sandy loam and generally ripen seeds, from which, or from young cuttings in sand, they are usually propagated. They flower from July to August, and 1. *microphylla* all the year.

TEPHROSIA *grandiflora, stricta* and *pallens*, are Cape shrubs, flowering in May and June, which may be treated like *Indigofera*.

ASTRAGALUS *tomentosus* and *Chinensis*, an Egyptian and Chinese perennial, which flower in June and July, grow in sandy loam, and are increased by dividing at the root or by seeds.

DALEA *Lagopus*, a Mexican annual, which flowers in

October and November, and may be raised on heat, and
then transplanted into small pots of light rich soil for
the green-house.

PSORALEA *pinnata*, A. R. 474.

P. *odoratissima*.

P. *verrucosa*.

P. *aculeata*, B. R. 146, B. M. 2158.

P. *bracteata*, B. M. 446.

P. *spicata*, A. R. 411.

P. *aphylla*, B. M. 1727, B. C. 221.

P. *multicaulis*.

P. *tenuifolia*.

P. *decumbens*, B. C. 282.

P. *hirta*.

P. *Stachyos*.

P. *repens*.

P. *bituminosa*.

P. *glandulosa*, B. M. 990.

P. *pedunculata*, B. R. 228.

P. *palæstina*.

P. *americana*.

P. *capitata*.

Most of these are Cape shrubs flowering from May
to August: they grow in loam and peat, and are increased by young cuttings or seeds which they ripen
freely.

LOTUS *glaucus* is a Madeira biennial, and L. *anthyl-
loides* a Cape shrub, both flowering in July and August;
and L. *australis*, B. M. 1365, is a New Holland shrub
flowering from May to September. Culture as in
Psoralea.

L. *jacobæus*, B.M. 79, is a Cape Verd shrub, valuable
as flowering all the year.

LOTUS *creticus,* and L. *hirsutus,* B. M. 336, are shrubs from the South of Europe, which flower from July to August. These require peat and loam as a soil.

L. *odoratus,* B. M. 1233, is a Barbary perennial, which grows in sandy soil and is increased by dividing at the root.

DORYCNIUM *monspeliense,* a shrub from the South of Europe, which flowers from July to September, grows in sandy loam, and is increased by young cuttings in sand under a bell-glass.

MEDICAGO *arborea,* a frame shrub from Italy, which grows in any light soil, and is increased by cuttings or seeds which it produces in abundance.

SCHRANKIA *uncinata,* a North American perennial of common culture in any loamy soil.

ACACIA *verticillata,* B. M. 110, B. C. 535.

A. *juniperina,* B. C. 398.

A. *acicularis.*

A. *sulcata.*

A. *suaveolens,* B. C. 730.

A. *glaucescens.*

A. *floribunda.*

A. *linifolia,* A. R. 394.

A. *stricta,* A. R. 53, B. C. 99, B. M. 1121.

A. *longifolia,* A. R. 207, B. C. 678.

A. *falcata.*

A. *melanoxylon,* B. M. 1659, B. C. 630.

A. *Sophoræ.*

A. *marginata.*

A. *myrtifolia,* B. M. 302.

A. *hispidula.*

A. *decipiens,* B. M. 1745.

A. *biflora.*

A. *armata,* B. M. 1653, B. C. 49.

Acacia *alata*, B. R. 396.

A. *pulchella*, B. C. 212.

A. *ciliata.*

A. *nigricans*, B. M. 2188, B. C. 313.

A. *pubescens*, B. M. 1263.

A. *lophantha*, A. R. 563, B. R. 361, B.C. 716, B. M. 2108.

A. *glandulosa.*

A. *decurrens*, B. R. 371.

A. *reticulata.*

A. *Caffra.*

A. *discolor*, A. R. 235, B. C. 601.

These are valuable shrubs for the green-house, many of them (see Part I. p. 84.) flowering from Christmas to April. They are chiefly from Australasia and the Cape; but one or two are from America, viz. A. *glandulosa* and *brachyloba*, the latter a perennial from Illinois. All the species thrive in loam, peat, and sand; most of them root by cuttings of the young wood under a bell-glass in sand, and plunged in heat. But some kinds do not root readily from cuttings, and these Mr. Sweet informs us " may be increased by taking off roots, as large pieces as can be spared, and planting them in the same kind of soil as the old plants, when they should be plunged under a hand-glass in a little bottom heat. Most of the kinds might be propagated by that means." *Bot. Cult.* 126.

Gleditschia *sinensis*, a frame shrub from China, grows in sandy loam, and is increased by young cuttings under a bell in sand.

Ceratonia *siliqua*, the St. John's Bread or Carob-tree, A. R. 567, a shrub from Spain and the Levant, which grows in light soil, and roots in sand under a bell-glass. Its seeds (beans) in Spain are used as horse-corn, and even by the common people in times of dearth.

ROSACEÆ.

ACÆNA *latebrosa*, a Cape perennial flowering from April to June, of common culture.

AMYGDALUS *orientalis*, silvery-leaved Almond, a Levant shrub flowering in March and April, and propagated by cuttings or grafting on A. *nana*.

PRUNUS *japonica*, Japan Plum, B. R. 27, a low tree flowering from March to May.

P. *prostrata*, birch-leaved Plum, B. R. 136, a frame shrub from Crete, flowering in April and May.

These plants grow in loamy soil, and ripened cuttings root freely in sand under a hand-glass.

CHRYSOBALANUS *oblongifolius*, American cocoa Plum, a low tree from Georgia. Culture as in *Prunus*.

MESPILUS *lævis, glabra*, and *indica*, B. M. 1726, are Chinese shrubs, which flower from April to July.

M. *arbutifolia* is a handsome shrub from California, flowering in July and August.

M. *japonica*, the Loquat, is a low tree from Japan, flowering in October, and bearing an edible berry or pome.

The culture of these plants is the same as in *Prunus*.

CYDONIA *japonica*, B. M. 622, and *speciosa*, are handsome-flowering frame shrubs from Japan; the first flowers in May and June, and the other all the year. Culture as in *Prunus*.

ROSA *berberifolia*, berberry-leaved Rose, a frame shrub from Persia, flowering in June and July.

R. *Banksiæ*, Lady Banks's Rose, B. M. 1954, B. R.

397, a Chinese evergreen shrub flowering in June and July.

Rosa *sinica,* three-leaved China Rose, a low frame shrub from China, flowering from May to July.

R. *semperflorens,* ever-blowing Rose, single and double, B. M. 284, a shrub from China, which in a warm green-house flowers all the year.

R. *Lauranceana,* Miss Laurance's Rose, B. M. 1762, B. R. 538, a shrub from the Mauritius, flowering all the year.

R. *odorata,* sweet Chinese Rose, flowering from February to August.

These beautiful shrubs grow in light rich soil, and are increased by cuttings of ripened wood in the same soil, with the addition of sand.

Rubus *rosæfolius,* rose-leaved Bramble, double and single varieties, B. M. 1783, flowering from October to April.

R. *pinnatus,* pinnate Bramble, a Cape shrub flowering in June and July.

These plants require similar soil and treatment to the Rose.

Poterium *caudatum* is a Canary shrub which flowers from January to April.

P. *spinosum,* a Levant shrub flowering from April to August.

These plants are of no beauty : they grow in any rich light soil, and are increased by young cuttings in the same soil under a hand-glass.

Cliffortia *cuneata, ilicifolia, tridentata, ruscifolia, cinerea pulchella, crenata, ericæfolia, obcordata, trifoliata,* and *sarmentosa,* are Cape shrubs, flowering from May to August; the whole of them grow in loam and peat, and are increased by young cuttings in sand. Only two or three are worth culture, as forming neat bushes and being evergreens.

SALICAREÆ.

LYTRHUM *alatum*, B. M. 1812, a North American perennial flowering from May to November, growing in loamy soil, and increased by cuttings or division at the root.

CUPHEA *viscosissima* and *procumbens*, B. R. 182, an American biennial and annual, which may be raised in heat and transplanted into small pots. They grow in any light rich soil, and increase by cuttings or seeds.

MELASTOMACEÆ.

RHEXIA *ciliosa*, a frame perennial from Carolina, which grows in loam and peat, and is increased by dividing at the root: it flowers from June to August.

MYRTEACEÆ.

BÆCKIA *frutescens*, a Chinese shrub flowering from October to December.

B. *virgata*, A. R. 598, and *densifolia*, shrubs from New South Wales, flowering from August to October. These are pretty and very hardy plants; they grow in loam and peat, and are increased by young cuttings in sand under a bell-glass.

LEPTOSPERMUM *scoparium*, New Zealand Tea, A. R. 622.

L. *flavescens*.

L. *attenuatum*.

L. *lanigerum*.

L. *pubescens*.

LEPTOSPERMUM *grandifolium*, B. M. 1810, B. C. 701.
L. *parvifolium*.
L. *stellatum*.
L. *arachnoideum*.
L. *juniperinum*.
L. *baccatum*.
L. *imbricatum*.
L. *triloculare*.
L. *ambiguum*.

Shrubs from New South Wales, flowering in May, June, and July; grown in loam and peat, and increased by cuttings of young wood in pots of sand.

FABRICIA *myrtifolia* and *lævigata*, B. M. 1304, New Holland shrubs, flowering in May and June, of common culture in loam and peat.

METROSIDEROS *hispida*, B. C. 106.
M. *floribunda*.
M. *costata*.
M. *glomulifera*.
M. *angustifolia*.
M. *marginata*.
M. *linearis*.
M. *pinifolia*.
M. *viminalis*.
M. *saligna*, B. M. 1821.
M. *lanceolata*, B. M. 260.
M. *speciosa*, B. M. 1761, B. C. 285.

Shrubs from New South Wales, and one, M. *angustifolia*, from the Cape: they flower at different periods from March to June; and M. *lanceolata* flowers from November till June.

EUGENIA *elliptica*, B. M. 1872, a shrub from New South Wales, flowering from May to October.

Myrtus *communis*, common Myrtle, a well-known shrub from the South of Europe, of which there are the following varieties in cultivation.

M. *communis romana*, broad-leaved Myrtle.

M. *communis tarentina*, box-leaved Myrtle.

M. *communis italica*, upright Myrtle.

M. *communis bœtica*, orange-leaved Myrtle.

M. *communis lusitanica*, Portugal Myrtle.

M. *communis belgica*, broad-leaved Dutch Myrtle.

M. *communis mucronata*, rosemary-leaved Myrtle.

M. *tomentosa*, woolly-leaved Myrtle, B. M. 250, a Chinese shrub flowering in June and July. All the species are of easy culture in sandy loam and leaf-mould.

Eucalyptus *robusta*.

E. *rostrata*.

E. *pilularis*.

E. *tereticornis*.

E. *resinifera*, A. R. 400.

E. *marginata*.

E. *capitellata*.

E. *saligna*.

E. *globulus*.

E. *botryoides*.

E. *hæmastoma*.

E. *piperita*.

E. *obliqua*.

E. *corymbosa*.

E. *paniculata*.

E. *cornuta*.

These are Australasian shrubs, but a few of which have flowered in this country: they all grow in loamy soil with a little leaf-mould, and are increased by young cuttings in sand under a bell-glass.

PUNICA *nana*, dwarf Pomegranate, B. M. 634, a West Indian shrub flowering from July to September.

This shrub grows in any light soil, and is easily increased by half-ripened cuttings in sandy loam under a hand-glass.

MELALEUCA *viridiflora*.

M. *paludosa*.

M. *globifera*.

M. *diosmifolia*, A. R. 476.

M. *styphiloides*.

M. *genistifolia*.

M. *striata*.

M. *thymoides*.

M. *squamea*, B. R. 477, B. C. 412.

M. *nodosa*.

M. *ericifolia*.

M. *armillaris*, A. R. 175.

M. *uncinata*.

M. *scabra*.

M. *pulchella*, B. C. 200.

M. *thymifolia*, B. M. 1868, B. C. 439.

M. *decussata*.

M. *fulgens*, B. R. 103, B. C. 378.

M. *linariifolia*.

M. *hypericifolia*, A. R. 200, B. C. 199.

M. *squarrosa*, B. M. 1935.

M. *calycina*.

M. *densa*.

These are all Australasian shrubs flowering from June or July to September, and growing in loam and peat. Cuttings of the young wood root freely in sand under bell-glasses.

TRISTANIA *ncreifolia*, B. M. 1058, B. C. 157, *law ina*

and *conferta*, are shrubs from New South Wales, flowering from June to September, growing in loam and peat, and increased by half-ripened cuttings in the same soil under a hand-glass.

CALOTHAMNUS *quadrifida*, B. M. 1506, B. C. 737, *villosa*, B. C. 92, and *gracilis*, are New Holland shrubs flowering from June to September. They may be treated as *Tristania*.

BEAUFORTIA *decussata*, B. R. 18, and *sparsa*, are New Holland shrubs, which may be treated like *Calothamnus* or *Tristania*.

CUCURBITACEÆ.

LUFFA *fœtida*, B. M. 1638, a frame annual from India, flowering from June to October. It should be raised in heat and then transplanted into any light rich soil.

TRICOSANTHES *anguina*, common Snake-gourd, B. M. 722, a frame annual from China, flowering in May and June.

T. *cucumerina*, cucumber-like Snake-gourd, a frame annual from the East Indies, flowering in June and July, and succeeded by cucumber-like fruit.

These plants thrive best when treated like the common Cucumber.

CUCURBITA *aurantia*, and *Citrullus*, and CUCUMIS of its various species, though frame plants, belong more properly to culinary than ornamental gardening.

LOASEÆ.

BARTONIA *ornata*, B. M. 1487, and *nuda*, are biennials from Missouri, flowering from July to September: they

grow in any light rich soil, and are increased by cuttings
or seeds.

MENTZELIA *oligosperma*, B. M. 1760, a perennial from
Louisiana, flowering in May and June; growing in
loamy soil, and increased as are other perennials.

ONAGRAREÆ.

ŒNOTHERA *nocturna* and *villosa*, Cape biennials which
flower from April to June.

Œ. *rosea*, B. M. 347, a perennial from Peru, which
flowers from May to August.

These are showy plants, of easy culture in sandy
loam, and increased by seeds, cuttings, or dividing the
root.

GAURA *mutabilis*, B. M. 388, a frame biennial from
South America, which flowers in July and August, and
is easily cultivated and increased in light soil.

EPILOBIUM *villosum*, a Cape perennial, flowering in
July and August, of easy culture, and increased in any
loamy sandy soil.

HALORAGIS *Cercodia*, a New Zealand shrub, flowering
from April to September.

JUSSIEUA *grandiflora*, a perennial from Carolina,
flowering from July to October, and growing in sandy
loam.

MONTINIA *caryophyllacea* is a Cape shrub flowering
in July: it grows in sandy loam, and is readily increased
by cuttings beginning to ripen.

FICOIDEÆ.

GLINUS *lotoides*, an annual from the South of Europe,

which may be raised on heat in any light soil : it flowers in July and seeds freely.

TETRAGONIA *expansa, crystallina, fruticosa, decumbens, tetrapteris, spicata, herbacea,* and *echinata,* are Cape shrubs and herbs of no beauty : the first species, a native of New Zealand, is cultivated as a Spinage : all of them are succulent in their leaves, grow in sandy soil, and are increased by cuttings or seeds with the greatest ease.

Of *the genus* MESEMBRYANTHEMUM there are nearly a hundred species in addition to those we have given below; but as few of these have found their way into the commercial gardens of this country, we have preferred a selection of those which may be purchased at the principal nurseries. Whoever wishes to see the finest collection of this family of plants in Europe, must visit the gardens of the Prince Salm-Dyck near Anhalt. From the *Hortus Dyckensis* we have selected some of the following names as preferable to those in common use here.

MESEMBRYANTHEMUM	MESEMBRYANTHEMUM
cordifolium	*lævigatum*
limpidum	*diminutum*
Aitonis	*pugioniforme*
anatomicum	*elongatum*
tortuosum	*capitatum*
varians	*diversifolium*
expansum	*corniculatum*
humifusum	*tricolorum*
Tripolium	*acinaciforme*
sessiliflorum	*rubro-cinctum*
lanceolatum	*dimidiatum*
loreum	*æquilaterale*
dubium	*crassifolium*
pubescens	*virescens*
perviride	*filamentosum*
edule	*spectabile*
subulatum	*australe*

MESEMBRYANTHEMUM

conspicuum
rubricaule
serrulatum
sarmentosum
glaucescens
minimum
cylindricum
quadrifidum
bibracteatum
minutum
bifidum
compactum
teretifolium
teretiusculum
obcordellum
truncatellum
fibuliforme
nuciforme
testiculare
fissum
obtusum
calamiforme
subulatum
bellidiflorum
acutum
pisiforme
moniliforme
digitatum
robustum
luteo-viride
cruciatum
gibbosum
difforme
heterophyllum
semicylindricum
glomeratum
mutabile
inclaudens
marginatum
spinosum
decumbens

MESEMBRYANTHEMUM

falciforme
pulchellum
microphyllum
deltoides
caulescens
muricatum
mucronatum
falcatum
confertum
stipulaceum
Haworthii
maculatum
hirtellum
emarginatum
tenuifolium
versicolor
deflexum
pulverulentum
floribundum
echinatum
hirsutum
stelligerum
striatum
hispidum
micans
polyanthon
retroflexum
scabrum
barbatum
subglobosum
brevifolium
speciosum
nodiflorum
caducum
apetalum
tuberosum
splendens
spiniferum
flexuosum
grossum
micranthon

(110)

MESEMBRYANTHEMUM

viridiflorum
reflexum
nitidum
fastigiatum
canaliculatum
ciliatum
villosum
junceum
brachiatum
geniculiflorum
anceps
virgatum
stellatum
filicaule
maximum
compressum
curvifolium
heteropetalum
serratum
aduncum
spiniforme
lucerum
depressum
adscendens
longum
albidum
præpingue
linguæforme
canum
scalpratum
rufescens
latum
dolabriforme
bicolorum
horizontale
verruculatum
graniforme
noctiflorum

MESEMBRYANTHEMUM

aurantium
aureum
glaucum
strictum
mo
stramineum
purpureo-croceum
coccineum
cymbiforme
Helianthoides
pinnatifidum
crystallinum
glabrum
pomeridianum
foliosum
radiatum
rigidum
viride
tenellum
parviflorum
geminatum
bracteatum
tumidulum
umbellatum
multiflorum
forficatum
uncinatum
perfoliatum
curtum
vaginatum
imbricatum
vulpinum
denticulatum
ramulosum
murinum
tigrinum
caninum
felinum

The culture of *Mesembryanthemum* is of the easiest kind, and has been already detailed in describing a few select species for a choice assortment in a small green-house, Part I. p. 106.

AIZOON *canariense*, A. R. 201, *glinoides*, *hispanicum*, and *lanceolatum*, are Cape and South of Europe succulents of no beauty and of the easiest possible culture.

REAUMURIA *hypericoides*, a frame shrub from Syria, flowering from July to October, which grows in sandy soil, and is increased by cuttings in the same soil.

SEMPERVIVEÆ.

LAROCHEA *falcata*, *coccinea*, B. M. 495, *cymosa*, *flava*, and *odoratissima*, A. R. 26, Cape Succulents which grow in sandy soil or lime rubbish, requiring little water and propagating freely by cuttings, which after being taken off and prepared should be left a day to dry up their wounds before being planted.

CRASSULA
   ramosa
   arborescens, B. M. 384
   tetragona
   lactea, B. M. 1771, B. C. 735
   scabra
   pellucida
   cordata, B.C. 359
   moschata
   glabra
   retroflexa
   expansa
   perfoliata
   mollis
   acutifolia
   obliqua
   ciliata
   undulata, B. C. 584
   columnaris
   imbricata

CRASSULA
   punctata
   marginalis
   spathulata
   cultrata, B. M. 1940
   obvallata
   nudicalis
   tomentosa
   Cotyledon
   lineolata
   orbicularis
   centauroides, B. M. 1765
   dichotoma
   glomerata
   Alooides
   capitella
   sparsa
   diffusa
   versicolor, B.R. 320, B.C. 433
   coccinea, B. M. 495, B.C. 486

The culture of this tribe of Succulents has been already given at length, Part I. p. 105.

SEPTAS *capensis*, A. R. 99, *globifera*, B. M. 1472, and *umbella*, are Cape perennial succulents, which may be treated as *Larochea*.

COTYLEDON *orbiculata, ovata*, B. M. 321, *oblonga, ramosissima, fascicularis, hemisphærica, spuria*, and *cæspitosa*, are Cape succulents of little beauty. See Part I. p. 105.

SEDUM *divaricatum, spinosum*, and *nudum*, are succulents of the easiest culture, but without being worth cultivating.

SEMPERVIVUM *arboreum*, B. R. 99, and S. *a. variegatum*, are curious little succulent trees, both of which may be admitted in a select green-house. When they flower it is from December to March.

S. *canariense, glutinosum*, B. R. 278, B. M. 1963, *glandulosum, ciliatum*, B. M. 1978, *tortuosum*, B. M. 296, *stellatum, africanum, dodrantale*, and *monanthos*, B. M. 93, are succulents from Madeira, Teneriffe, and other islands, of the easiest culture.

PORTULACEÆ.

PORTULACARIA *afra*, African Purslane-tree, a handsome little succulent tree with a reddish-tinged bark. It is of the easiest culture in sandy loam, or lime rubbish and a little leaf-mould.

LIMEUM *africanum*, a Cape perennial, a succulent-leaved plant of easy culture.

ANACAMPSEROS *rotundifolia*, B. C. 591, *arachnoides*, B. M. 1368, *rubens, filamentosa*, B. M. 1367, and *lanceolata*, are Cape succulents of no beauty, but of the easiest culture.

These plants may be propagated by leaves. Take them off when full grown and hardened a little, and with the whole of their petioles, then let them dry a few days to heal the wound; after that plant them up to the base of the leaf in sandy soil, and they will root and shoot out young plants from the lower end of the petiole.

CACTEÆ.

CACTUS *flagelliformis*, B. M. 17, a creeping succulent from Peru, without leaves, but with fine deep red flowers.

C. *Opuntia*, Indian Fig, B. M. 1577, a prickly succulent from the South of Europe, which has no beauty in its flowers, but produces a fruit sometimes eaten by connoisseurs.

These plants are of the easiest culture in dry rich soil.

HYDRANGEA *hortensis*, B. M. 438, a frame undershrub from China which produces very showy flowers from April to October. It requires a light rich soil, large pots, to be often shifted and parted, and abundance of water when in a growing state.

H. *quercifolia*, B. M. 975, a frame shrub from Florida, of much less beauty than the other, but requiring similar treatment to cause it to flower freely.

SAXIFRAGEÆ.

SAXIFRAGA *sarmentosa*, B. M. 92, is a perennial, with succulent leaves and very productive of red stolones, from China: it grows in sandy soil, and its stolones and plantlets hang down from the pot in a singular and pleasing manner. It flowers in June and July, and is very hardy.

There a number of frame plants which belong to this genus, but they are chiefly natives of Britain and Switzerland, and are in correct language frame Alpines. They do not therefore belong to the green-house.

TIARELLA *biternata*, a frame perennial from Carolina, which flowers in May and June, and grows readily in loam with a little peat.

CUNONIACEÆ.

CUNONIA *capensis*, a Cape shrub.

CALLICOMA *serratifolia*, A. R. 556, a shrub from New South Wales, flowering in May and June.

ARALIACEÆ.

CUSSONIA *thyrsiflora* and *spicata*, Cape shrubs, which have not yet flowered in this country.

CAPRIFOLEÆ.

LONICERA *sempervirens minor*, B. R. 556, B. M. 1753.

L. *flava*, B. M. 1318, B. C. 338.

L. *japonica*, B. R. 70.

L. *flexuosa*, B. R. 712.

The two first of these are American, and the two last Chinese, twining frame shrubs, of great beauty and fragrance, grown in loam and peat, and increased by cuttings.

VIBURNUM *rugosum*, B. R. 376, B.M. 2082, a Canary evergreen shrub, of great beauty and easy culture in loam and peat.

UMBELLIFEREÆ.

HYDROCOTYLE *asiatica, repanda,* and *villosa,* are perennials from the Cape and North America, of common culture.

Bupleurum *scorzoneræfolium* and *nudum* are Cape perennials of no beauty, but of the easiest culture.

B. *coriaceum*, *fruticescens*, *canescens*, *spinosum*, and *difforme*, are South of Europe and African shrubs of no beauty, but of easy culture in sandy loam.

Conium *rigens* and *africanum* are perennials, Cape Hemlock's ugly plants, as easily grown as the common weeds of that name.

Selinum *decipiens* is a low Madeira shrub of no beauty, but of easy culture.

Peucedanum *aureum*, a biennial from the Canaries, of the usual culture.

Crithmum *latifolium*, Canary Samphire, a biennial, which grows in sand and lime rubbish, and is increased by seeds or cuttings.

Bubon *macedonicum*, *Galbanum*, *lævigatum* and *gummiferum*, are ugly plants from Greece and the Cape: they are grown in sandy soil, and increased by seeds and sometimes by division of the root. B. *Galbanum* affords the gum of that name by inspissating the juice of the root.

Hermas *depauperata* and *gigantea* are Cape perennials of the usual culture.

Arctopus *echinatus*, a Cape perennial which may be classed with *Hermas*.

CORYMBIFEREÆ.

Liatris *odoratissima*, A.R. 638, a Carolina perennial, of the usual culture in loamy soil.

Mikania *scandens*, a climbing perennial from North America, of the easiest culture.

Eupatorium *urticæfolium*, a frame perennial from North America of no beauty.

Ageratum *cœlestinum*, B. M. 1730, B. C. 623, a low

blue flowering shrub of little beauty and the easiest culture.

Stevia *salicifolia,* a frame perennial from Mexico, an ugly plant, which grows in loam and peat.

Cephalophora *glauca,* a frame perennial from Chili, of the usual culture.

Hymenopappus *tenuifolius,* a frame biennial from Louisiana, of the usual culture.

Melananthera *hastata,* and *hastata pandurata,* perennials from North America which grow in loam and peat, and are increased by cuttings or dividing at the root: their leaves are their only recommendation.

Marshallia *lanceolata* and *latifolia,* frame perennials from Carolina, of common culture.

Bidens *heterophylla,* a perennial from Mexico, of common culture in sandy loam.

Cacalia *papillaris.*

C. *Anteuphorbium.*

C. *Kleinia.*

C. *Ficoides.*

C. *carnosa.*

C. *repens.*

C. *Haworthi.*

C. *articulata.*

C. *tomentosa.*

C. *appendiculata.*

C. *scandens.*

These are low spongy succulents with yellow flowers, by no means desirable for the green-house. They may be treated like other succulents.

Piqueria *trinervia,* a Mexican perennial of no beauty.

Chrysocoma *comaurea,* B. M. 1972.

C. *cernua.*

CHRYSOCOMA *ciliaris.*

C. *scabra.*

C. *denticulata.*

Ugly yellow-flowering chamomile-like undershrubs, easily grown and increased in sandy soil.

TARCHONANTHUS *camphoratus,* B.C. 382, a low spongy Cape shrub of no beauty; it grows in sandy loam, and is readily increased by cuttings.

HUMEA *elegans,* a biennial from New South Wales, of some beauty, and of the usual culture.

IXODIA *achillæoides,* B. M. 1534, a low shrub from New Holland, of the usual culture.

ATHANASIA *capitata.*

A. *pubescens.*

A. *dentata.*

A. *trifurcata.*

A. *virgata.*

A. *tomentosa.*

A. *filiformis.*

A. *crithmifolia.*

A. *parviflora.*

A. *pectinata.*

Low spongy Cape shrubs of easy culture in sandy loam.

BALSAMITA *ageratifolia,* a low spongy shrub from Candia, which grows in sandy loam, and is increased by cuttings in the same soil.

TANACETUM *linifolium* and *suffruticosum,* low spongy shrubby Tansy plants from the Cape, of the easiest culture in light soil.

ARTEMISIA *judaica, valentina, subcanescens, tenuifolia, argentea, palmata, chamæmifolia,* and *indica,* are frame and green-house spongy shrubs and half herbs, of no beauty, and growing in light soil.

GNAPHALIUM *eximium*, A. R. 654.

G. *crispum.*

G. *arboreum.*

G. *grandiflorum,* A. R. 489.

G. *fruticans,* B. M. 1802.

G. *congestum,* B. R. 243.

G. *patulum.*

G. *discolorum.*

G. *cephalotes.*

G. *fastigiatum.*

G. *milleflorum.*

G. *diosmæfolium.*

G. *ericoides,* B. M. 435.

G. *teretifolia.*

G. *ignescens.*

G. *crassifolium.*

G. *maritimum.*

G. *dasyanthum.*

G. *orientale.*

G. *cymosum.*

G. *rutilans.*

G. *apiculatum,* B. R. 240.

G. *odoratissimum.*

G. *fœtidum,* B. M. 1987.

G. *helianthemifolium.*

G. *squarrosum.*

G. *purpureum.*

G. *declinatum.*

G. *glomeratum.*

These are low spongy white-leaved undershrubs, chiefly from the Cape, and valued for the durability of their flowers after being gathered. They grow in sandy peat, and are increased by cuttings generally in the same soil, or with an additional quantity of sand.

ELICHRYSUM *vestitum.*

E. *spirale,* A. R. 262.

E. *speciosissimum,* A. R. 51.

E. *dealbatum.*

E. *fulgidum,* B. M. 414.

E. *variegatum.*

E. *proliferum,* B. R. 21, B. C. 8.

E. *canescens,* B. M. 420.

F.. *argenteum.*

E. *retortum.*

E. *humile,* A. R. C. ic.

E. *sesamoides,* B. M. 425.

E. *fasciculatum,* A. R. 242.

E. ——— *album,* A. R. 279.

E. ——— *rubrum,* A. R. 650.

E. *rigidum,* A. R. 387.

E. *ericoides.*

E. *Stæhelina,* A. R. 428.

E. *fragrans,* A. R. C. ic.

E. *herbaceum,* A. R. 487.

E. *paniculatum.*

This is a handsome family of plants intermediate between shrubs and herbs. They have mostly white hairy or woolly leaves, and their flowers are exceedingly brilliant, generally yellow and orange, and preserve their colours a long time in a dried state. They grow in sandy peat, the pots being well drained and not over-large. Cuttings root in sand on heat, but uncovered, as the confinement of a bell is apt to damp off the leaves. They are almost all Cape plants, a d flower at various periods, but chiefly from May to August.

BACCHARIS *angustifolia, ivæfolia* and *neriifolia,* American and a Cape undershrub of similar habits to *Elichrysum,* and requiring similar treatment. They are of no beauty, and only curious on account of their white woolly leaves.

Conyza *camphorata.*

C. *candida.*

C. *verbascifolia.*

C. *hirsuta.*

C. *Gouani.*

C. *fœtida.*

C. *sordida.*

C. *saxatilis.*

C. *rupestris.*

C. *sericea.*

C. *inuloides.*

C. *incisa.*

C. *virgata.*

C. *rugosa.*

Low spongy whitish shrubs from different parts of the South of Europe, Africa, and America, of no beauty. They grow in loam and peat, and young cuttings root in sand under a bell-glass.

Erigeron *glaucum,* B. R. 10, a low spongy shrub which flowers great part of the year, but has little else to recommend it. It grows freely in loam and peat, and cuttings root readily under a hand-glass.

Tussilago *fragrans,* B. M. 1388, a frame perennial from Italy, valued for its fragrance. It grows in any soil, and increases itself like the common *Tussilago.* One plant will perfume a whole green-house.

Senecio *reclinatus.*

S. *purpureus.*

S. *erubescens.*

S. *divaricatus.*

S. *cinerascens.*

S. *hastatus.*

SENECIO *venustus.*

S. *elegans.*

S. ——*flore pleno,* B. M. 238.

S. *speciosus,* B. R. 41.

S. *rosmarinifolius.*

S. *asper.*

S. *rigescens.*

S. *lanceus.*

S. *longifolius.*

S. *halimifolius.*

S. *ilicifolius.*

S. *rigidus.*

These are low spongy shrubs, perennials or annuals, mostly from the Cape, and generally flowering late in the season. None of them deserve a place in the greenhouse, excepting S. *purpureus, cinerascens, elegans,* and *speciosus,* which have brilliant-petalled red, blue or purple flowers, and S. *elegans pleno* flowers all the year. They grow in loam and peat, and cuttings root with the greatest ease in sandy soil uncovered.

ASTER *reflexus,* B. M. 884.

A. *tomentosus,* A. R. 61.

A. *sericeus.*

A. *Cymbalariæ.*

A. *lyratus,* B. M. 1509.

A. *argophyllus,* B. M. 1563.

A. *angustifolius.*

A. *villosus.*

A. *obtusatus.*

A. *fruticulosus.*

A. *filifolius.*

A. *tenellus,* B. M. 33.

These are plants with similar habits to *Senecio*; few of them are of any beauty, and all of them are easily grown in loam and peat, and rooted in sandy soil uncovered. They are mostly from the Cape, and flower late in the season.

CINERARIA *gelifolia.*

C. *parviflora.*

C. *aurita,* B. M. 1786.

C. *lactea.*

C. *cruenta,* B. M. 406.

C. *hybrida.*

C. *populifolia.*

C. *lobata.*

C. *malvæfolia.*

C. *petasites,* B. M. 1536.

C. *maritima.*

C. *linifolia.*

C. *humifusa.*

C. *viscosa.*

C. *lanata,* B. M. 53.

C. *amelloides,* B. M. 249.

These are spongy undershrubs and herbs nearly allied to *Aster* in habit, and of the same culture and propagation. C. *cruenta* has very dark purple leaves. C. *amelloides* and *aurita* have purple flowers; and as plants of similar colours are generally scarce, they may be admitted: but few of the others deserve a place in a select green-house.

INULA *viscosa,* a perennial from the South of Europe, of no beauty, and of the easiest culture.

GRINDELIA *glutinosa,* B. R. 187.

G. *inuloides,* B. M. 248.

G. *squarrosa,* B. M. 1706.

These are spongy shrubs, and the last a perennial, which grow in any light rich soil, and are increased by cuttings or seeds.

PODOLEPIS *rugata* and *acuminata,* B. M. 956. New Holland perennials of common culture.

XIMENESIA *encelioides,* a Mexican biennial of common culture in light rich soil.

HELENIUM *quadridentatum,* a frame perennial from Louisiana, of no beauty, and of the easiest culture.

TAGETES *lucida,* B. M. 740, an American sweet-scented perennial, which flowers from July to November: it grows in sandy loam, and is increased as other perennials are.

LEYSERA *gnaphalodes,* and *squarrosa,* Cape undershrubs, of no beauty and easy culture.

CHRYSANTHEMUM *pinnatifidum,* a Madeira shrub of the easiest culture.

Pyrethrum fœniculaceum, B. M. 272.

P. *crithmifolium.*

P. *anethifolium.*

P. *frutescens.*

P. *coronopifolium.*

P. *grandiflorum.*

P. *speciosum.*

Low spongy ugly shrubs from Teneriffe and the Canaries, of the easiest culture and propagation.

MATRICARIA *capensis,* a Cape biennial of the most simple treatment and without beauty.

LIDBECKIA *pectinata,* a Cape shrub, flowering in May and June, and valued for its silvery leaves. It is of the easiest culture in light rich soil.

COTULA *coronopifolia,* a Cape annual of the easiest culture.

Anthemis globosa, a perennial from the South of Europe.

A. *artemisiæfolia,* a Chinese perennial, of whose numerous and beautiful varieties, and culture in pots and the green-house, we have already treated (P. I. page 126).

ACHILLEA *ægyptiaca*, a Levant perennial of the easiest culture.

AMELLUS *Lychnitis*, a hardy undershrub from the Cape, of no beauty, but easily cultivated and increased in sandy loam.

VERBESINA *alata*, B. M. 1716, and *serrata*, South American perennials of no beauty, and the easiest culture.

ZALUZANIA *triloba*, a frame perennial from Mexico, of the usual culture.

BUPHTHALMUM *frutescens, arborescens, sericeum*, B.M. 1836, and *maritimum*, are spongy plants between shrubs and herbs, of little elegance and easily cultivated.

HELIANTHUS *dentatus*, a Mexican perennial, of no merit but facility of cultivation.

RUDBECKIA *lævigata*, a frame perennial from Carolina, ugly, and of easy culture.

COSMEA *lutea*, B. M. 1689, and *bipinnata*, B. M. 1535, Mexican herbs of easy culture.

COREOPSIS *ferulæfolia*, B. M. 2059, a frame Mexican perennial of the easiest culture.

OSMITES *camphorina*, a low spongy shrub from the Cape, which smells like camphor. It grows and increases as freely as any plants of this Order.

PALLASIA *halimifolia*, a Peruvian shrub of easy culture.

SCLEROCARPUS *africanus*, an annual from Guinea, ugly, and easy of culture.

CULLUMIA *ciliaris*, B. M. 2059, B. C. 302, *setosa*, and *squarrosa*, low Cape shrubs, spongy-wooded, like all the *Corymbifereæ*, and of the easiest culture.

BERCKHEYA *cynaroides, obovata, incana, cuneata, palmata, grandiflora*, B. M. 1844, *uniflora* and *cernua*, Cape shrubs of common culture in loam and peat.

DIDELTA *carnosum* and *spinosum*, Cape shrubs, spongy, and which may be treated like *Berckheya*.

GORTERIA *personata,* a Cape annual of no beauty and easy culture.

GAZANIA *rigens,* B. M. 90, *pavonia,* B. R. 35, and *subulata,* Cape perennials of common culture.

ARCTOTHECA *repens,* a Cape perennial of the easiest culture.

SPHENOGYNE *crithmifolia, scariosa, abrotanifolia, dentata,* and *odorata,* Cape shrubs, spongy, and easily cultivated and increased.

CENTAUREA *hyssopifolia, intybacea, cineraria, ragusina,* B. M. 494, *argentea, sempervirens, canariensis, ferox,* and *ægyptiaca.*

Spongy shrubs and perennials, chiefly frame plants, and all of easy culture in sandy loam, and increased by cuttings in the same soil uncovered.

ALCINA *perfoliata,* a Mexican annual of the easiest culture in any rich light soil.

CALENDULA *graminifolia,* B. R. 289.

C. *Tragus,* B. M. 408.

C. ⸺ *flaccida,* B. R. 28.

C. *viscosa,* A. R. 412.

C. *dentata,* A. R. C. ic.

C. *oppositifolia.*

C. *fruticosa.*

C. *chrysanthemifolia,* B. R. 40.

C. *arborescens.*

These are very low spongy Cape shrubs, but of considerable beauty: they grow in loam and peat, mostly flower in May, and are increased by cuttings of ripened wood without being covered or placed on heat.

ARCTOTIS *acaulis,* B. R. 122.

A. *tricolor,* B. R. 131.

A. *undulata.*

A. *grandiflora.*

ARCTOTIS *glaucophylla.*

A. *plantaginea.*

A. *argentea.*

A. *rosea.*

A. *decumbens.*

A. *angustifolia.*

A. *flaccida.*

A. *decurrens.*

A. *melanocicla.*

A. *reptans.*

A. *auriculata.*

A. *fastuosa.*

A. *spinulosa.*

A. *maculata,* B. R. 130.

A. *aspera,* B. R. 34.

A. *aureola,* B. R. 32.

A. *bicolor.*

These are Cape perennials, and very low spongy shrubs, some of them of considerable beauty. They are all treated as *Calendula.*

OSTEOSPERMUM *spinosum.*

O. *spinescens.*

O. *pisiferum,* B. C. 470.

O. *moniliferum.*

O. *ilicifolium.*

O. *rigidum.*

O. *cœruleum.*

O. *polygaloides.*

These are low spongy shrubs, which grow well in any light rich soil, and cuttings root freely under a hand-glass.

OTHONNA *pinnata,* B. M. 768.

O. *pectinata,* B. M. 306.

O. *Athanasiæ.*

OTHONNA *abrotanifolia*, B. R. 108.

O. *retrofracta.*

O. *coronopifolia.*

O. *cheirifolia*, B. R. 266.

O. *crassifolia.*

O. *denticulata.*

O. *heterophylla.*

O. *Lingua.*

O. *filicaulis.*

O. *bulbosa.*

O. *perfoliata*, B. M. 1312.

O. *parviflora.*

O. *ericoides.*

O. *tenuissima.*

O. *arborescens.*

O. *cuculioides.*

Low spongy Cape shrubs and perennials, which may be treated as *Osteospermum.*

HIPPIA *frutescens*, B. M. 1855, a low Cape shrub, which may be treated like *Osteospermum.*

GYMNOSTYLES *anthemifolia,* a South American annual which flowers from April to December: it grows in light rich soil, and seeds freely.

ERIOCEPHALUS *africanus*, B. M. 833, and *racemosus,* low Cape shrubs, of the usual culture of the *Corymbiferæ.*

ELEPHANTOPUS *carolinianus* and *tomentosus,* American perennials of the easiest culture.

ŒDERA *prolifera*, B. M. 1637, a low spongy shrub from the Cape, of the easiest culture.

STŒBE *æthiopica* and *cinerea,* spongy Cape shrubs of the usual culture.

CASSINA *aurea,* a New Holland perennial of common culture.

FRANSERIA *artemisioides* and *ambrosioides*, low spongy South American shrubs of easy culture.

RUBIACEÆ.

CRUCIANELLA *pubescens* and *maritima*, the first a perennial from Candia, the other an undershrub of a herbaceous nature from the South of Europe : they grow in sandy loam, and are easily increased by cuttings of the roots or shoots.

RUBIA *lucida*, *fruticosa*, and *angustifolia*, low shrubs of a herbaceous nature, which may be cultivated and increased like *Crucianella*.

BOUVARDIA *triphylla* and *versicolor*, B. R. 245, American shrubs of easy culture.

ZIERA *Smithii*, B. M. 1395, a shrub from New South Wales, of easy culture in loam and peat, and cuttings root in the same soil with a little sand added, and covered with a hand-glass.

MUSSÆNDA *pubescens*, B. M. 2099, a shrub from China, of common culture.

PINCKNEYA *pubens*, a frame shrub from Georgia, of easy culture.

GARDENIA *radicans*, B. R. 73.

G. *Thunbergia*, B. M. 1004.

G. *Rothmannia*.

G. *florida*, B. R. 449, and varieties.

G. *spinosa*.

China and Cape shrubs, beautiful plants ; and G. *radicans* and *florida* are very odoriferous and much in demand on that account. One plant of either will scent a whole room in the evening if ever so crowded. G. *florida* is properly a stove plant, and indeed all the species require to be placed in the warmest part of the green-house; but they flower best on a frame with a little moist heat. Cuttings root in sand under a glass in bottom heat.

Serissa *fœtida* and *flore pleno*, B. M. 361, a Japan shrub which grows in loam and peat; and cuttings root readily in sand under a hand-glass.

Pæderia *fœtida*, a China shrub which may be treated as *Serissa*.

Plocama *pendula*, a shrub from the Canaries of easy culture.

Phyllis *Nobla*, a Canary shrub which may be treated like *Serissa*.

Anthospermum *æthiopicum*, a Cape shrub of easy culture and propagation in loam and peat.

CYNAROCEPHALEÆ.

Cnicus *Casauboni, afer*, and *diacanthus*, ugly perennials from the South of Europe of common culture.

Cynara *horrida* and *glomerata*, Cape and Madeira artichokes of common culture.

Stokesia *cyanea*, a Carolina perennial, rather pretty, of common culture.

Stobæa *pinnata*, B. M. 1788, a Cape spongy shrub which flowers all the year: it grows in any rich light soil, and cuttings root freely under a hand-glass.

Stæhelina *arborescens* and *Chamæpense*, shrubs from Candia of easy culture and hardy.

Pteronia *camphorata, stricta, flexicaulis, oppositifolia*, and *scariosa*, Cape shrubs of little beauty which grow freely in loam and peat, and cuttings root under a hand-glass.

Sphæranthus *africanus*, a Cape shrub of easy culture in any light rich soil.

CENTAUREA *hyssopifolia, spinosa, ragusina, cineraria, argentea, sempervirens, canariensis, ferox,* and *ægyptiaca,* perennials, chiefly frame plants, and some of them approaching to undershrubs, of the easiest culture and propagation.

DIPSACEÆ.

MORINA *persica,* a perennial, from Persia, of the usual culture.

SCABIOSA *rigida, attenuata, africana, cretica* and *lyrata,* suffruticose plants and herbs, from the Cape and the South of Europe, of common culture in light rich soil.

GENTIANEÆ.

EXACUM *viscosum,* a biennial from the Canaries, of the easiest culture.

VILLARSIA *lacunosa, sarmentosa,* B. M. 1328, *parnassifolia,* B. M. 1029, and *ovata,* Cape, New Holland, and North American aquatics which require to be grown in water, the pots being placed only a few inches below the surface; the soil used may be peat and loam, and the plants are increased by dividing at the root or by seeds.

LOGANIA *latifolia* and *floribunda,* A. R. 520, New Holland shrubs, which thrive in sandy loam and peat, and ripened cuttings root in sand under a hand-glass.

CHIRONIA *jasminoides,* B. R. 197, B. C. 27.

C. *lychnoides.*

C. *linoides,* B. M. 511.

C. *baccifera,* B. M. 233.

C. *angustifolia,* B. M. 818.

C. *frutescens,* B. M. 37.

C. *decussata,* B. M. 707.

These are spongy-wooded low Cape shrubs, pretty plants, with handsome flowers, but of short lives, and therefore require to be frequently renewed by cuttings. They grow in peat with a little loam, and cuttings in the same soil, under a bell-glass, root readily.

ERYTHRÆA *maritima*, a perennial from the South of Europe, of the usual culture in sandy loam.

CICHORACEÆ.

SONCHUS *fruticosus, pinnatus, lævigatus, lyratus,* and *radicatus,* low spongy shrubs from Madeira and the Canaries, of no beauty, and growing in any light soil rather rich.

PRENANTHES *spinosa,* a frame undershrub from Barbary, of no beauty, but of the easiest culture.

HIERACIUM *fruticosum,* a Madeira undershrub of the easiest culture.

CREPIS *rigens* and *filiformis,* a perennial and biennial from the Azores and Madeira, which grow in light soil, and are readily increased.

ANDRYALA *cheiranthifolia, pinnatifida, crithmifolia,* and *ragusina,* perennials and biennials from Madeira and the Archipelago, which grow in light loam, and cuttings root in the same soil under a hand-glass.

CICHORIUM *spinosum,* a frame biennial from Candia, which grows in any light soil.

CAMPANULACEÆ.

LIGHTFOOTIA *oxycoccoides* and *subulata,* a Cape shrub and perennial, which grow in light loam, and are increased by cuttings in the same soil under a hand-glass.

CAMPANULA *gracilis,* B. M. 691, *peregrina,* B. M.

1257, *cernua capensis*, B. M. 782, *mollis*, B. M. 404, *saxatilis, laciniata, aurea*, B. R. 57, *fruticosa, prismato-carpus, pentagonia*, B.R. 56, and *perfoliata*, shrubs, per-ennials, and annuals of the easiest culture in any light rich soil.

Roella *ciliata*, B. M. 378, *squarrosa, decurrens* and *muscosa*, a Cape shrub, annuals and a perennial of com-mon culture.

Phyteuma *pinnata*, a perennial from Candia of com-mon culture in light soil.

Trachelium *diffusum*, a Cape undershrub, which may be treated as *Phyteuma*.

Lobelia *simplex*.

L. *linearis*.

L. *pinifolia*, A. R. 240.

L. *unidentata*, B. M. 1484.

L. *salicifolia*, B. M. 1325.

L. *gigantea*, B. M. 1325.

L. *bellidifolia*.

L. *triquetra*.

L. *secunda*.

L. *alata*.

L. *fulgens*, A. R. 659, B. R. 165.

L. *splendens*, B. R. 60.

L. *cardinalis*, B. M. 320.

L. *debilis*.

L. *gracilis*, B. M. 741.

L. *minuta*.

L. *Erinus*, B. M. 901.

L. *erinoides*.

L. *pubescens*.

L. *bicolor*, B. M. 514.

L. *lutea*, B. M. 1319.

L. *variifolia*, B. M. 1692.

L. *hirsuta*.

LOBELIA *coronopifolia*, B. M. 644.

L. *ilicifolia*, B. M. 1896.

L. *crenata*.

L. *Speculum*, B. M. 1499.

These are chiefly Cape and American perennials, some of them with very showy flowers: they grow in light rich soil, and cuttings root with ease, covered by a hand-glass. The culture of the three frame species, *L. fulgens, splendens,* and *cardinalis,* is the same as that of the *Dahlia*.

CYPHIA *volubilis* and *bulbosa,* Cape perennials, which grow in loam and peat, and are increased by cuttings under a hand-glass.

CANARINA *campanula*, B. M. 444, B. C. 376, a perennial which grows in any light rich soil, and roots readily under a hand-glass.

MICHAUXIA *campanuloides*, B. M. 219, a biennial of the easiest culture.

STYLIDEÆ.

STYLIDIUM *graminifolium*, B. R. 90, B. C. 385, *fruticosum*, B. C. 171, and *scandens*, New Holland perennials and a shrub which grow in sandy loam and leaf-mould, and are increased by cuttings divided at the roots, or by seeds.

RHODORACEÆ.

AZALEA *indica*, B. M. 1480, B. C. 275, an Indian shrub of great beauty and fragrance, which grows in sandy turfy peat well drained, and requires a warm part of the green-house.

KALMIA *hirsuta*, B. M. 138, a frame shrub from

North America, which grows in peat soil, and is best increased by layers.

RHODODENDRON *dauricum*, B. M. 636, B. R. 194, B.C. 605, and R. *Chamæcistus*, B.M. 488, frame shrubs from Siberia, which may be treated like *Kalmia*.

BEJARIA *racemosa*, a handsome shrub from Florida, which grows in peat soil, and is increased by cuttings in sand under a bell on heat, or better by layers.

ERICEÆ.

BLÆRIA *ericoides, articulata, purpurea, muscosa,* and *ciliaris,* heath-like Cape shrubs, which may be treated like *Erica.*

CYRILLA *caroliniana*, a Carolina shrub, which may be treated like *Erica.*

VACCINIUM *meridionale, myrtifolium, nitidum,* B. R. 480, *n. decumbens*, B. M. 1550, and *myrsinites,* low evergreen creeping American shrubs, which grow in peat soil, and are increased by cuttings of the tender tops, or by layers.

ANDROMEDA *japonica*, a Japan shrub which may be treated like *Vaccinium.*

ENKIANTHUS *quinqueflora*, B. M. 1649, a China shrub of the culture of *Erica.*

ARBUTUS *canariensis*, B. M. 1577, and *phillyreæfolia,* shrubs from Canary and Peru, which grow in peat soil, and are best increased by layers.

CLETHRA *arborea*, B.M. 1057, and *a. minor*, low trees from Madeira, to be treated as *Arbutus.*

HUDSONIA *ericoides*, a frame shrub from America, which may be treated like *Erica.*

EMPETRUM *album*, a Portugal heath-like shrub, to be treated as *Erica.*

(135)

The ERICA family are almost as numerous as the *Geraniaceæ*, and like them they admit of interminable increase and variety, by the production of seed-bearing mules. It is supposed by several of the best informed botanists, that many of what we call species have been so originated by nature in the heathery surfaces of the Cape; and it is certain that a number have been so produced in the green-houses about London. From these, as given in the Catalogues of Lee, Andrews, Cormack, Page, Donn, Sweet, and others, we might have greatly increased our list; but we prefer a selection, and have rejected some names because the plants no longer exist; others, because we consider them as only varieties; and some, as not to be purchased at the public nurseries.

ERICA

Sebana, A. H. vol. 1. B. C. 23.
furfurosa, A. H. vol. 1.
monadelpha, B. M. 1370.
follicularis, A. H. vol. 1.
Banksii, Ib. vol. 1.
leucanthera.
socciflora, Ib. vol. 1.
Petiveriana.
Plukenetiana, Ib. vol. 1.
penicilliflora.
penicillata, Ib. vol. 2.
melastoma, Ib. vol. 1. B. C. 333.
melanthera.
flexuosa, A. H. vol. 1.
imbricata, Ib. vol. 2.
villosa, Ib. vol. 3.
velleriflora, Ib. vol. 1.
tiaræflora, Ib. vol. 3.
vagans.
 β alba.
spumosa, B. C. 566.
sexfaria.

ERICA

Brunoides.
umbellata, A. H. vol. 2.
latifolia, Ib. vol. 2.
nudiflora.
carnea, B. M. 11.
mediterranea, Ib. 471.
multiflora, A. H. vol. 2.
gelida, Ib. vol. 2. B. C. 699.
discolor, Ib. vol. 1.
coccinea, Ib. vol. 2.
grandiflora, B.M.189. B.C.498.
cruenta, A. H. vol. i.
verticillata, Ib. vol. 1. B. C.145.
ignescens, Ib. vol. 2.
cylindrica, Ib. C. ic.
curviflora, Ib. vol. 1.
conspicua, Ib. vol. 2.
splendens.
Ewerana, Ib. vol. 2.
speciosa, Ib. vol. 2. B. C. 575.
mammosa, A.H. vol.1. B.C.125.
densifolia.

ERICA

clavæflora, A. H. vol. 2.

Patersoni, Ib. vol. 1.

fascicularis, Ib. vol. 1.

perspicua.

corifolia.

glauca, B. M. 580.

triflora.

versicolor, A.H. vol. 1. B.C.208.

flammea, A. H. vol. 2.

procera.

squamosa, A. H. vol. 3.

luchneæflora, Ib. vol 3.

cerinthoides, B. M. 220.

costata, A. H. vol. 1.

Aitoniana, B. M. 429. B. C 144.

Lawsoni, B. M. 1720 B. C. 488.

Niveni, A. H. vol. 2.

tubiflora, Ib. vol. 1.

simpliciflora.

spuria, Ib. vol. 1.

laxa, Ib. vol. 3.

tenuiflora, Ib. vol. 3.

fistulæflora, Ib. vol. 3.

elegans, B. M. 966. B. C. 185.

pyrolæflora.

Hyacinthoides, A. H. vol. 3.

aristata, A. H. vol. 3. B. C. 73.
B. M. 1249.

tricolor, A. H. vol. 3.

— *aristata minor*, A. H.

andromedæflora, B.C.521 B.M.
1250.

ovata, A. H. vol. 2.

acuminata, Ib. vol. 3.

lucida, Ib. vol. 2.

vestita.

α *alba*, A.H. vol. 1. B.C.243.

β *incarnata*, Ib. vol. 2.

γ *purpurea*, Ib. vol.1 B.C.217.

ERICA

δ *rosea*, A.H. vol. 2.

ε *fulgida*, Ib. vol. 2.

ζ *coccinea*, Ib. vol 1. B. C. 55.

η *lutea*, Ib. vol. 1.

pellucida, Ib. vol. 3. B. C. 276.

serratifolia, Ib. vol. 1.

Sparrmanni, Ib. vol. 3.

Archeriana, Ib vol. 2.

Hibbertiana, Ib. vol. 3 B. M.
1758.

Massoni, B. M. 356.

radiata, A. H. vol. 1.

rosea, Ib. vol. 2.

princeps, Ib. vol. 2.

tetragona, Ib. vol. 3.

blanda, Ib. vol 3.

transparens, B. C. 177.

Linnæana, A.H. vol.2. B C.102.

longifolia, B. M. 706.

coccinea, A. H. vol. 1.

exsurgens, Ib. vol. 1.

prægnans, Ib. C. ic.

jasminiflora, Ib vol. 1.

ferruginea, Ib. vol. 3.

ventricosa, B. M. 350. B C.431.

ampullacca, B. M. 303. B.C.508.

glandulosa.

erubescens, A. H. vol 3.

elata, Ib. vol. 2.

purpurea, Ib. vol. 2. B. C. 703.

aurea, Ib. vol. 2.

pinea.

Monsoniæ, B. M. 1915.

Halicacaba, A. H. vol. 2.

lanuginosa, Ib. vol. 3.

retorta, B. M. 362.

inflata.

Leeana, A. H. vol. 1. B. C.298.

viridis, Ib. vol. 2. B. C. 233.

(137)

ERICA
comosa
α rubra.
β alba.
ardens, B. R. 115. B. C. 47.
nitida.
globosa, A. H. vol. 3.
Tetralix.
β alba.
filamentosa, B. R. 6. B. C. 395.
margaritacea, A. H. vol. 1.
empetrifolia, B. M. 447.
lateralis, A. H. vol. 1.
canescens, Ib. vol 2.
eriocephala.
odorata, A.H. vol. 3. B.M. 1399.
    B. C. 633.
Thunbergii, B. M. 1214. B. C.
    277.
exigua.
parviflora.
florida, B. C. 234.
nigrita, A. H. vol. 1. B. C. 34.
buccans, B. M. 358.
oppositifolia.
petiolata, Ib. vol. 1.
laxifolia, Ib. vol. 1.
patens, Ib. vol. 3.
canaliculata, Ib. vol. 3.
flava, Ib. vol. 2. B. M. 1815.
Blandfordiana, Ib. vol. 3. B. M.
    1793. B. C. 115.
cinerea.
β alba.
stricta, A. H. vol. 2.
decora, Ib. vol. 3.
horizontalis, Ib. vol. 2.
physodes, B. M. 443. B. C. 223.
obliqua, A. H. vol. 1.
pilulifera.
catervæfolia.

ERICA
pyramidalis, B. M.366. B.C.319.
propendens, A. H. vol. 2. B. C.
    63. B. M. 2140.
tardiflora.
Solandri, A. H. vol. 2.
empetroides, Ib. vol. 2.
acuta, Ib vol. 2.
Bergiana, Ib. vol. 2.
barbata. Ib. vol. 2.
Muscari, Ib. vol. 1.
Passerina.
daphnæflora, B. C. 543.
albens, B M. 440. B. C. 95.
ciliaris, B. M. 484.
lutea, A. H. vol. 1. B.C. 64.
glutinosa.
urceolaris.
flexicaulis, A. H. vol. 2.
fastigiata, Ib. vol. 1. B. C. 207.
    B. M. 2084.
cubica, Ib. vol. 1.
cchiiflora, Ib. vol. 3. B.C. 364.
viscaria, B. C. 726.
formosa.
α alba.
β rubra.
hirtiflora, B. M. 481.
hispidula.
malleolaris.
cordata, A. H. vol. 3.
primuloides, B. M. 1548. B. C.
    715.
depressa, A. H. vol. 2.
australis, Ib. vol. 3.
palustris, Ib. vol. 2. B. C. 4.
pulchella, B.C. 307.
cernua.
pendula.
Lambertiana, A. H. vol. 2.
rubens, B. C. 557.

(138)

ERICA

scariosa, B. C. 477.
bracteolaris.
incarnata, A.H. C. ic.
gnaphalodes.
lanceolata.
Broadleyana, Ib. C. ic.
calyculata.
emarginata, Ib. C.ic.
fimbriata, Ib. C. ic.
cephalotes.
Caffra, Ib. C.ic. B. C. 196.
Savilii, Ib. C. ic. B.C. 96.
axillaris.
pectinifolia.
incurva, A. H. C.ic.
pallens, Ib. C. ic.
humifusa.
vesicularis.
carduifolia.
auricularis.
tomentosa.
helianthemifolia.
glomiflora.
tubercularis.
lavandulæfolia.
corydalis.
azaleæfolia.
verniciflora.
lyrigera.
salax.
cumuliflora.
genistæfolia.
periplocæfolia.
bicolor.
thymifolia, A. II. vol. 2.
retroflexa.
planifolia.
marifolia, A. II. vol. 1.
phylicoides.
scoparia.

ERICA

campanulata, B.C. 184.
conferta, A. H. vol. 2.
racemosa.
tenella, A. H. vol.2.
viridipurpurea.
regerminans.
imbecilla.
gracilis, L. T.
gracilis, B. C. 244.
amœna.
ramentacea, A. H. vol. 1. B. C.
446.
strigosa.
racemifera, A. H. vol. 3.
persoluta, B. M. 342.
paniculata.
arborea.
β stylosa.
pubescens, B. M. 480. B. C. 167.
absinthoides.
setacea, A. H. vol. 1.
fragrans, Ib. vol. 2. B. M. 2181.
B. C. 288.
hyssopifolia.
pulviniformis.
brachialis.
festa.
parilis.
borboniæflora.
fallax.
notæflora.
doliiformis.
sceptriformis.
Swainsoniana, A. H. C.ic.
Sainsburyana, Ib. C. ic.
moschata, Ib. C. ic. B. C.614.
magnifica, Ib. C. ic.
Shannoni, Ib. C. ic. B. C. 168.
rugosa, Ib. C. ic.
densa, Ib. C. ic.

ERICA
hispida, A. H. C. ic.
siccifolia.
oxycoccifolia.
cinerascens.
coarctata.
fugax.
divaricata.
squarrosa.
tragulifera.
pavettæflora.
pusilla.
intervallaris.
pubigera.
curvirostris.
turbiniflora.
embothriifolia.
capax.
lasciva.
diotæflora.
holosericea.
brevifolia.

ERICA
chlamydiflora.
velitaris.
cyrillæflora.
fabrilis.
selaginifolia.
filiformis.
stagnalis.
succiflora.
palliiflora.
lasciva.
nutans.
pannosa.
cuspidigera.
modesta.
tegulæfolia.
diosmæfolia.
stylosa.
turgida.
vestiflua.
squamæflora.
cæsia.

The culture and management of this genus has been already given at length.

EPACRIDEÆ.

ANDERSONIA *sprengelioides*, B. M. 1645, B. C. 263, a New Holland shrub which grows in sandy loam and leaf-mould, and must be sparingly watered: young cuttings treated like those of *Erica* root freely.

EPACRIS *purpurascens*, B. M. 844, B. C. 237.

E. *pulchella*, B. M. 1170, B. C. 194.

E. *grandiflora*, B. M. 982, B. C. 21.

E. *obtusifolia*, B. C. 292, and *exserta*.

Australasian shrubs of considerable beauty, and valuable as flowering early in spring: they grow in rough, turfy, sandy peat soil, the pots frequently shifted, as the mass of roots round the insides of the pots is apt to be destroyed during the hot sunshine of summer. Young cuttings grow in sand under bell-glasses, and succeed best when put in in autumn or very early in spring.

MONOTOCA *elliptica* and *lineata*, New Holland shrubs, which grow in sandy loam and peat well drained, and are increased by young cuttings in sand under a bell-glass.

LEUCOPOGON *lanceolatus*, A. R. 287, *ericoides, amplexicaulis* and *juniperinus*, B. C. 447, shrubs from New South Wales, which may be treated like *Monotoca*.

STENANTHERA *pinifolia*, B. R. 218, B. C. 228, a beautiful plant, which may be treated like *Epacris*.

ASTROLOMA *humifusum*, B. M. 1439, a low bushy New Holland shrub, with scarlet flowers which come out from May to October : it grows in sandy loam and peat, and young cuttings root in sand under a bell.

STYPHELIA *longifolia*, B. R. 24.

S. *viridiflora*, A. R. 312.

S. *triflora*, B. M. 1297.

S. *tubiflora*.

Beautiful New Holland shrubs, which flower from April to June, and may be treated as *Astroloma*.

MYRSINEÆ.

ARDISIA *excelsa* and *crenata*, a Madeira tree and shrub, which grow in loamy soil, and are increased by half-ripened cuttings in sand under a hand-glass.

MYRSINE *africana, retusa*, B. C. 409, *samara*, and *melanophelos*, Cape shrubs which grow in loam and peat, and cuttings root under cover in sand.

SAPOTEÆ.

SIDEROXYLON *inerme*, a Cape shrub which may be treated like *Ardisia*.

SERSALISIA *sericea*, a New Holland shrub of the same nature as *Sideroxylon*.

EBENACEÆ.

ROYENA *lucida*.   R. *hirsuta*.
   villosa.   *angustifolia*.
   pallens.   *ambigua*.
   glabra.   *polyandra*.
   pubescens, B. R. 500.

Cape shrubs of some beauty of foliage, which grow in loam and peat; and ripened cuttings root in sand under a hand-glass.

VISNEA *mocanera*, a shrub from the Canaries, which may be treated like *Royena*.

HOPEA *tinctoria*, a Carolina shrub, very hardy, which grows freely in loam and peat, and cuttings strike root in the same soil.

DIOSPYRUS *Kaki*, a Japan shrub or low tree, which grows in light rich loam, and is increased by grafting on any of the hardy species, or by ripened cuttings in sand.

OLEINEÆ.

LIGUSTRUM *lucidum*, the Wax-tree, a Chinese shrub which grows in sandy loam, and is readily increased by cuttings.

OLEA *europea*, and several varieties.

O. *capensis*, B. R. 613.

Olea *undulata.*

O. *verrucosa.*

O. *americana.*

O. *excelsa.*

O. *fragrans,* B. M. 1552.

Evergreen shrubs with small white flowers, which grow in sandy loam, and increase freely by cuttings in the same soil.

Notelæa *longifolia* and *ligustrina,* New Holland shrubs, which grow in loam and peat, and are increased by cuttings in sand under a bell-glass.

JASMINEÆ.

Jasminum *gracile,* A. R. 127, B. R. 606.

J. *glaucum.*

J. *azoricum,* B. R. 89.

J. *odoratissimum,* B. M. 285.

J. *revolutum,* B. R. 178, B. M. 1731.

J. *grandiflorum,* B. R. 91.

Oriental shrubs, climbers with fragrant flowers, whose culture has been already noticed (Part I. p. 35 and 93).

VERBENACEÆ.

Stachytarpheta *orubica,* a biennial from Panama, of easy culture in any light rich soil.

Callicarpa *americana,* an American shrub, which grows in loam and peat, and ripened cuttings root under a hand-glass.

Selago *corymbosa, canescens, diffusa, ovata,* B. M. 186, *polygaloides, spuria, fasciculata,* B. R. 184, and *lucida,* Cape shrubs, which grow in loam and peat, and cuttings root in the same soil under a glass.

HEBENSTREITIA *dentata*, B. M. 483, *integrifolia*, A. R. 252, *ciliata*, *spicata*, *crinoides*, and *cordata*, are Cape shrubs, an annual and perennial of easy culture in loam and peat, and young cuttings root in sandy loam under a hand-glass.

CLERODENDRUM *trichotomum* and *tomentosum*, B. M. 1518, shrubs from Japan and New South Wales, which grow in loam and peat, and young cuttings root in sandy loam under a hand-glass, on heat.

VITEX *Agnus-castus*, a frame shrub from Sicily, and *incisa*, B. M. 364, a Chinese shrub, both of which grow in light loamy soil, and cuttings root in sand under a glass.

SPIELMANNIA *africana*, B. R. 1899, a Cape shrub flowering from February to November, of no great beauty, but which grows in any light rich soil, and is increased by cuttings in the same soil covered with a hand-glass.

ZAPANIA *nodiflora*, an American shrub of the same culture as *Spielmannia*.

ALOYSIA *citriodora*, B. M. 367, a lemon-scented deciduous shrub, very hardy, and which may be increased by cuttings either of the young or old wood.

VERBENA *Aubletia*, B. M. 308, B. R. 294, an American biennial of the easiest culture.

ASCLEPIADEÆ.

PERIPLOCA *lævigata*, a shrub from the Canaries, of the easiest culture in any light soil.

SECAMONE *ægyptiaca*, a shrub from Egypt, which grows in any light soil, and increases by cuttings in the same soil.

MICROLOMA *sagittatum*, a Cape shrub which grows

in peat and loam, and cuttings root in sand under a bell-glass.

Cynanchum *obtusifolium* and *pilosum*, B. R. 111, Cape shrubs, handsome climbers, of the culture of *Microloma*.

Xysmalobium *undulatum*, a Cape shrub which may be treated like *Microloma*.

Gomphocarpus *arborescens*, *crispus* and *fruticosus*, B. M. 1628, Cape shrubs to be treated as *Microloma*.

Asclepias *parviflora* and *linaria*, perennials from North America of the easiest culture.

Marsdenia *suaveolens*, B. R. 489, a shrub from New South Wales, of the easiest culture in loam and peat.

APOCYNEÆ.

Arduinia *bispinosa*, B. C. 387, a neat little Cape shrub, which grows well in loam and peat, and cuttings root in sand under a bell-glass.

Nerium *Oleander* and *odorum*, B. R. 74, with their varieties, handsome-flowering shrubs, the culture of which has been already given.

Echites *difformis*, a Carolina shrub of easy culture.

BIGNONIACEÆ.

Cobæa *scandens*, B. M. 851, a Mexican climber of rapid growth and great show, of easy culture in any light rich soil. It is best increased by seeds.

Bignonia *grandiflora*, B. M. 1938, B. R. 418, *australis*, B. M. 865, and *cærulea*, climbing shrubs which grow in any light rich soil, and increase readily by cuttings.

Penstemon *campanulata*, B. M. 1878, a frame shrub from Mexico of the easiest culture.

(145)

Martynia *proboscidea*, B. M. 1056, and *longiflora*, an American and Cape annual of the usual culture.

POLEMONIACEÆ.

Vestia *lycioides*, B. R. 299, a Chili shrub of easy culture in loam and peat, and cuttings root freely under a bell-glass.

Ipomopsis *elegans*, a biennial from Carolina of the usual culture.

CONVOLVULACEÆ.

Convolvulus *pannifolius*, B. R. 222.
C. *canariensis*, B. M. 1228.
C. *farinosus*.
C. *suffruticosus*, B. R. 133.
C. *Hermanniæ*.
C. *erubescens*, B. M. 1067.
C. *bryoniæfolius*, B. M. 943.
C. *saxatilis*.
C. *Cneorum*, B. M. 459.
C. *linearis*, B. M. 289.
C. *floridus*.
C. *candicans*, B. M. 1603.

These are showy climbers from all parts of the world, which grow in loam and peat, and young cuttings root readily in sand under a bell-glass.

Ipomæa *dissecta, mutabilis*, B. R. 39, *Jalapa*, B. M. 1572, B. R. 342, and *pendula*, A. R. 613, B. R. 632, twiners which may be treated like *Convolvulus*.

Retzia *spicata*, a Cape shrub of the same culture as the two last genera.

Falkia *repens*, A. R. 257, a Cape perennial of the easiest culture.

(L)

DICHONDRA *repens*, a perennial from New South Wales of common culture.

CUSCUTA *chinensis*, a Chinese annual, a parasite which may be sown in peat-soil at the root of any other plant, and it will grow up and root into its branches, after which the root in the soil will die away.

BORAGINEÆ.

HELIOTROPIUM *peruvianum*, B. M. 141, and *corymbosum*, B. M. 1609, Peruvian shrubs, whose culture has been already given (Part I. page 93).

LITHOSPERMUM *distichum*, a Cuba perennial of the easiest culture.

ANCHUSA *capensis*, A. R. 336, a Cape biennial of the usual culture.

CYNOGLOSSUM *pictum*, B. M. 2134, and *lanatum*, herbaceous plants of the easiest culture.

ONOSMA *orientalis*, a Levant perennial of common culture.

TOURNEFORTIA *Messerschmidia*, a Canary shrub of the usual culture in light loamy soil.

ECHIUM *fruticosum*, B. R. 36.

E. *candicans*, B. R. 44.

E. *grandiflorum*, B. R. 124.

E. *ferocissimum*, A. R. 39.

E. *argenteum*, A. R. 154.

E. *lœvigatum*.

E. *glabrum*.

E. *fastuosum*.

E. *nervosum*.

E. *spicatum*.

Cape shrubs, spongy-wooded, and of little beauty of foliage, but with showy blue and red flowers. They grow in light loam, and increase easily by cuttings, and sometimes ripe seeds are produced.

Cordia *Patagonula,* a South American shrub which grows in loam and peat, and cuttings root in sandy loam.

SOLANACEÆ.

Verbascum *hæmorrhoidale, phlomoides* and *pinnatifidum,* herbaceous plants which grow in light soil, and are best increased by seeds.

Hyoscyamus *Senecionis, aureus,* B. M. 87, and *canariensis,* B. R. 180, spongy shrubs and a perennial of common culture.

Nicotiana *fruticosa, plumbaginifolia,* and *undulata,* B. M. 673, shrubs and herbs easily grown in light soil, and best increased by seeds, which they produce in abundance.

Atropa *frutescens,* a shrub from Spain of common culture.

Physalis *somnifera, aristata, flexuosa* and *peruviana,* B. M. 1068, shrubs and herbs which grow in light rich soil, and are increased by seeds or cuttings under a hand-glass. The berry of P. *peruviana* is edible and agreeable.

Solanum *Pseudo-Capsicum, crassifolium, laciniatum,* B. M. 349, B. C. 717, *radicans, bonariense, suffruticosum, Melongena, Campechiense, Carolinense, sodomeum, marginatum, tomentosum, Bahamense, giganteum, Milleri,* and *cordifolium,* shrubs and herbs of no beauty, and of easy culture in rich light soil, and increased by seeds or cuttings.

Capsicum *sinense,* a Cape undershrub which may be treated like *Solanum.*

Cestrum *Parqui,* B.M. 1770, and *venenatum,* shrubs

which grow in loam and peat, and cuttings root in sand under a hand-glass.

LYCIUM *rigidum, horridum* and *boerhaavefolium,* ugly shrubs from the Cape with fœtid smells, which may be treated like *Cestrum.*

CELSIA *Arcturus,* B.M. 1962, *cretica,* B. M. 964, and *lanceolata,* frame biennials from the South of Europe, of easy culture in light rich soil, and increased by cuttings or seeds. They are showy plants, and flower from June to October or later.

ALONSOA *acutifolia, incisifolia,* B. M. 417, and *line-aris,* B.M. 210, South American shrubs of easy culture.

ANTHOCERCIS *littorea,* B. R. 212, a curious but elegant little plant, which grows in loam and peat, and roots in sand under a bell.

SCROPHULARINEÆ.

CALCEOLARIA *pinnata,* B. M. 41, and *Fothergilli,* B. M. 348, an annual and perennial, showy and of easy culture in any light soil. They seed freely.

BUDDLEA *salvifolia* and *saligna,* Cape shrubs of easy culture in light rich soil, and increased by layers or cuttings.

BROWALLIA *demissa,* B. M. 1136, and *elata,* B. M. 34, American annuals of the easiest culture.

MAZUS *rugosus,* a frame annual from China of the usual culture.

CAPRARIA *lanceolata* and *undulata,* B.M. 1556, Cape shrubs which grow in loam and peat, and cuttings root freely in sand under a glass.

TEEDIA *lucida,* B. R. 210, and *pubescens,* B. R. 214, Cape biennials of the easiest culture.

HALLERIA *lucida,* B. M. 1744, a Cape shrub which may be treated like *Capraria.*

MIMULUS *glutinosus,* B. M. 354, a shrub from California, of the easiest culture on any light rich soil, and which flowers all the year.

M. *luteus,* a frame perennial of the usual culture.

DIGITALIS *canariensis,* B. R. 48, and *Sceptrum,* showy undershrubs from the Canaries and Madeira, of easy culture in sandy soil.

SCROPHULARIA *glabrata, frutescens, arguta, sambucifolia,* and *mellifera,* herbaceous plants of common culture.

MAURANDIA *semperflorens,* B. M. 460, and *antirrhinifolia,* B.M. 1643, climbing shrubs from Mexico of common culture. The first species is a very ornamental climber, and does not exceed due bounds in the greenhouse.

NEMESIA *chamædrifolia, fœtens* and *bicornis,* Cape herbaceous plants of the usual culture.

ANTIRRHINUM *Asarina,* B. M. 902, and *molle,* frame plants of easy culture in light rich soil.

LINARIA *triornithophora,* B. M. 525, *bipartita, tristis,* B. M. 74, *reticulata, alpina,* B. M. 207, *villosa,* and *origanifolia.* Frame herbaceous plants, which grow in loam and peat, and are increased by dividing at the root, or by cuttings.

PEDICULARIS *euphrasioides* and other species are properly alpine plants, though protected during winter by a frame.

MANULEA *fœtida,* A. R. 80, *villosa, Cheiranthus,* and *argentea,* are annuals and a biennial of the usual culture.

M. *pedunculata,* A. R. 84, *viscosa,* B. M. 217, *rubra* and *tomentosa,* B. M. 322, are Cape shrubs of easy culture in any light soil.

ERINUS *alpinus,* B. M. 310, *hispanicus,* and *fragrans,*

frame perennials, and a shrub of the same culture as *Manulea.*

DISANDRA *prostrata*, a perennial, a trailer, of the usual culture in light rich soil.

LABIATEÆ.

ROSMARINUS *chilensis*, a low evergreen shrub from Chili, which grows in sandy soil, and is readily increased by cuttings.

SALVIA *dentata.*

S. *angustifolia.*

S. *tibiæfolia.*

S. *serotina.*

S. *syriaca.*

S. *scabra.*

S. *runcinata.*

S. *rugosa.*

S. *nubia.*

S. *mexicana.*

S. *cæsia.*

S. *reptans.*

S. *formosa.*

S. *coccinea.*

S. *amarissima.*

S. *abyssinica.*

S. *canariensis.*

S. *aurita.*

S. *africana.*

S. *aurea*, B. M. 182.

S. *colorata.*

S. *chamædrioides*, B. M. 808, B. C. 576.

S. *paniculata.*

S. *spinosa.*

S. *tingitana.*

S. *odorata.*

Low shrubs, some of which are of considerable beauty : the best have been selected and their culture given before (Part I. pages 85 and 88).

COLLINSONIA *scabriuscula*, a perennial from Florida of common culture in peat soil

TEUCRIUM *nissolianum, trifidum, fruticans, latifolium,* B.M. 245, *Marum, regium, asiaticum, abutiloides, betonicum,* B. M. 1114, *massilense, heterophyllum, flavum, montanum, aureum, Polium, flavescens, gnaphalodes, capitatum, pumilum,* and *subspinosum.* Frame and green-house low evergreens of little beauty, but of easy culture, and very hardy. *T. Marum* is greedily devoured by cats, when they can come at it.

WESTRINGIA *rosmariniformis*, A. R. 214, and *Dampieri,* New Holland shrubs which grow in sandy loam and peat, and cuttings root under a bell-glass in sand.

SATUREJA *juliana, Thymbra,* and *græca*, frame perennials and low shrubs of easy culture in loam and peat.

THYMBRA *spicata*, and *verticillata*, shrubs from the South of Europe of the easiest culture.

LAVANDULA *Stœchas, viridis, dentata,* B. M. 401, *pinnata,* B. M. 400, *multifida,* and *abrotanoides,* evergreen shrubs which grow in any light rich soil, and are readily increased by cuttings in the same soil, or by seeds.

SIDERITIS *canariensis, candicans, syriaca, incana,* and *cristata,* low spongy shrubs, which may be treated as *Lavandula.*

BYSTROPOGON *plumosum, origanifolium, canariense,* and *punctuatum,* low shrubs from the Canaries, which grow and are increased in sandy peat and loam.

MENTHA *capensis*, a perennial of the easiest culture.

HYPTIS *radiata* and *persica*, a perennial and shrub of easy culture in sandy peat and loam, and increased by cuttings in the same soil.

STACHYS *coccinea*, B. M. 666, *spinosa*, *æthiopica*, and *rugosa*, shrubs and herbs of the easiest culture.

MARRUBIUM *africanum*, *Pseudo-Dictamnus*, and *acetabulosum*, frame and green-house perennials of the easiest culture.

PHLOMIS *Lychnites*, B. M. 999, a curious plant which grows in loam, sand, and peat, and is reared by cuttings on a little heat.

LEONOTIS *Leonurus*, B.M. 478, and *L. Leonotis*, Cape shrubs of easy culture.

ORIGANUM *Dictamnus*, B.M. 298, *ægyptiacum*, *siphyleum*, *Tournefortii*, A.R. 537, *smyrnæum*, *Majorana*, *salvifolium*, *majoranoides*, low spongy frame shrubs, which may be treated like *Salvia* or *Hyptis*.

THYMUS *mastichinus*, *cephalotus*, *villosus*, *Tragoriganum*, and *filiformis*, frame and green-house shrubs, which grow in any light soil (the poorer the more odoriferous the plants), and are increased by cuttings in the same soil.

ACYNOS *alpinus*, B. M. 2153, a frame biennial, which may be treated like *Thymus*.

CALAMINTHA *caroliniana*, B. M. 997, *cretica*, and *fruticosa*, low shrubs of the culture of *Thymus*.

DRACOCEPHALUM *canariense*, a well known odoriferous plant, grown in any light soil, and increased by cuttings or seeds.

OCYMUM *grandiflorum*, a low shrub of the easiest culture.

PLECTRANTHUS *fruticosus*, *parviflorus*, and *punctatus*, low shrubs or ligneous herbs, which may be treated like *Thymus*.

PROSTANTHERA *lasianthos*, B. R. 143, a beautiful New Holland plant of the usual culture, and propagation in loam and peat.

SCUTELLARIA *cretica*, a low shrub from Crete of the easiest culture.

PRASIUM *majus* and *minus*, low frame shrubs of easy culture in any light rich soil.

MYOPORINEÆ.

MYOPORUM *ellipticum*, A. R. 283, *acuminatum*, *parvifolium*, B. M. 1693, *tuberculatum*, *viscosum*, *debile*, B. M. 1830, *diffusum*, and *oppositifolium*, New Holland shrubs, which grow in loam and leaf-mould, and cuttings root in sandy loam under a bell-glass.

STENOCHILUS *glaber*, B. M. 1942, B. R. 572, a New Holland shrub of the same culture as *Myoporum*.

ACANTHACEÆ.

JUSTICIA *Adhatoda*, B. M. 861, *hyssopifolia*, and *orchioides*, shrubs which grow in loam and peat, and are easily increased by cuttings in sandy loam under a bell-glass.

RUELLIA *strepens*, *lactea*, and *biflora*, perennials of common culture.

LENTIBULAREÆ.

PINGUICULA *lutea*, B. R. 126, an annual from Carolina of the usual culture.

PRIMULACEÆ.

CYCLAMEN *persicum*, B. M. 44, a pretty little plant already noticed (Part I. page 107).

ANAGALLIS *fruticosa*, B. M. 831, *latifolia*, *Monelli*, and *linifolia*, pretty little biennials of easy culture and increase in peat-sand and vegetable mould.

SAMOLUS *littoralis,* an aquatic or marsh perennial from New South Wales, which is best grown in a pot of loam and peat immersed in water to the brim, but not more.

CORIS *monspeliensis,* B. M. 2131, a biennial of common culture.

GLOBULAREÆ.

GLOBULARIA *longifolia,* B. R. 685, *Alyssum,* and *spinosa,* shrubs which grow well in loamy soil, and cuttings root in sandy loam under a hand-glass.

PLUMBAGINEÆ.

PLUMBAGO *tristis,* a Cape shrub of easy culture in loamy soil.

STATICE *cordata, echioides, purpurata, pectinata, suffruticosa, monopetala, sinuata,* B. M. 71, *alata,* and *mucronata,* perennials whose culture has been already noticed. There are some frame species so hardy as to be considered mere alpines.

NYCTAGINEÆ.

OPERCULARIA *aspera,* a shrub from New South Wales, which grows in loam and peat, and cuttings root in sand under a hand-glass.

CRYPTOSPERMUM *Youngii,* a perennial of common culture.

MIRABILIS *dichotoma, jalapa* with several varieties, B. M. 371, *hybrida,* and *longiflora,* perennials with fusiform roots, which grow freely in any rich light soil, and seed abundantly, or may be preserved during winter like the Potatoe or *Dahlia.*

PISONIA *grandis*, a New Holland shrub, which may be treated as *Opercularia*.

AMARANTHACEÆ.

ACHYRANTHES *nivea*, a Canary shrub of common culture in any light soil.

ALTERNANTHERA *polygonoides*, a perennial of common culture.

PARONYCHIA *capitata* and *nivea*, perennials of the usual culture.

MOLLIA *diffusa*, an annual, and *aristata*, a shrub, both of the easiest culture.

HERNIARIA *polygonoides*, a low shrub from the South of Europe, which grows in any light soil, and increases freely by cuttings.

IRESINE *celosioides*, a frame perennial from America, of the easiest culture in peat soil.

CHENOPODEÆ.

SALICORNIA *arabica*, a shrub from Arabia, which grows freely in rich light soil not over-watered, and cuttings root in the same soil under a hand-glass.

CAMPHOROSMA *monspeliaca*, a low heath-like shrub, of no beauty, which grows in any sandy soil, or in lime rubbish, and is easily increased.

CHENOLEA *diffusa*, a silvery-leaved little plant which grows readily in a rich light soil, and cuttings root in sand under a bell-glass.

CHENOPODIUM *multifidum* and *anthelminticum*, plants of no beauty, but of the easiest culture in light soil.

BETA *patula*, a Madeira biennial, which grows in sandy loam and is increased by cuttings or seeds.

Salsola *prostrata*, a low shrub of the culture of *Salicornia*.

Anabasis *tamariscifolia*, a shrub from Spain of the easiest culture in light sandy soil.

Bosea *Yervamora*, a tree from the Canaries, which grows freely in loam and peat, and ripened cuttings root in sand under a bell-glass.

Galenia *africana*, a Cape shrub which grows in loam and peat, and cuttings root in sandy loam under a hand-glass.

Atriplex *glauca* and *albicans*, shrubs of similar culture to *Salicornia*.

Rhagodia *hastata*, a shrub from New South Wales which may be treated as *Chenolea*.

POLYGONEÆ.

Atraphaxis *spinosa* and *undulata*, a perennial and shrub of the usual culture in light sandy soil.

Rumex *Lunaria*, *giganteus* and *arifolius*, Docks of easy culture in light soil.

Polygonum *tinctorium*, a biennial from China of the usual culture.

Brunnichia *cirrhosa*, a climbing shrub from Carolina, which grows in loam and peat, and is increased by cuttings.

LAURINEÆ.

Laurus *Camphora*, *glauca*, *indica*, *fœtens*, *canariensis*, *Borbonia* and *carolinensis*, evergreen shrubs which grow in loam and peat, and cuttings root in sand under a bell-glass on a little moist heat.

PROTEACEÆ.

The culture of all the genera of this order is so nearly
the same, that we shall merely enumerate the species.
The culture has been given in the selection (Part I.
pages 85 and 86).

PETROPHYLA *pulchella,* B. M. 796, and *diversifolia.*

ISOPOGON *anethifolius,* B.M. 697, *formosus, anemoni-
folius,* and *trilobus.*

PROTEA *cynuroides,* B. M. 770.

P. *latifolia,* B. M. 1717.

P. *compacta.*

P. *longiflora.*

P. *speciosa,* B. M. 1183.

P. *obtusa,* A. R. 110.

P. *formosa,* B. M. 1713.

P. *melaleuca,* A. R. 103.

P. *Lepidocarpon,* A. R. 301.

P. *neriifolia,* B. R. 208.

P. *pulchella,* B. R. 20.

P. *patens,* A. R. 543.

P. *magnifica,* A. R. 438.

P. *longifolia,* B. R. 47.

P. *umbonalis,* A. R. 144.

P. *ligulæfolia,* A. R. 133.

P. *mellifera,* B. M. 346.

P. *grandiflora,* B. R. 569.

P. *minima.*

P. *Scolymus,* B. M. 698.

P. *mucronifolia,* B. M. 933.

P. *nana.*

P. *pendula.*

P. *tenax.*

P. *canaliculata,* A. R. 437.

PROTEA *acuminata*, B. M. 1694.

P. *acaulis*, B. M. 2065.

P. *acerosa*, B. R. 577.

P. *glaucophylla*.

P. *lævis*.

P. *scabra*.

P. *repens*.

P. *turbiniflora*.

P. *Scolopendrium*.

P. *cordata*, A. R. 289, B. M. 649.

P. *pumilis*, A. R. 532.

P. *amplexicaulis*.

LEUCOSPERMUM *candicans*, A. R. 294.

L. *formosum*, A. R. 17.

L. *lineare*.

L. *tollum*.

L. *medium*.

L. *ellipticum*.

L. *conocarpum*.

L. *grandiflorum*.

L. *puberum*.

L. *tomentosum*.

L. *parile*.

L. *hypophyllum*.

MIMETES *hirta, palustris, cucullata, divaricata, vac-cinifolia*, and *purpurea*.

SERRURIA *abrotanifolia*, A. R. 522.

S. *millefolia*, A. R. 337.

S. *pinnata*, A. R. 512.

S. *artemisiæfolia*, A. R. 264.

S. *arenaria*.

S. *cyanoides*.

S. *pedunculata*, A. R. 264.

S. *Niveni*, A. R. 349.

SERRURIA *ciliata.*

S. *phylicoides,* A. R. 507 and 4.

S. *æmula.*

S. *parilis,* A. R. 507.

S. *odorata,* A. R. 545.

S. *emarginata,* A. R. 536.

S. *glomerata.*

S. *decipiens.*

S. *Roxburgii.*

S. *Burmanni.*

S. *elongata.*

S. *triternata,* A. R. 447.

NIVENIA *Sceptrum, spathulata, spicata, media* and *crithmifolia,* A. R. 293.

SOROCEPHALUS *imberbis, spatalloides, tenuifolia, lanatus,* and *imbricatus,* A. R. 317.

SPATALLA *prolifera, incurva,* A. R. 429, and *Thunbergii.*

PERSOONIA *hirsuta,* B. C. 327, *linearis,* B. M. 760, *salicina,* and *lanceolata,* A. R. 74, B. C. 75.

GREVILLEA *sericea,* B. M. 862.

G. *linearis,* A. R. 272, B. C. 50.

G. *riparia.*

G. *arenaria.*

G. *acuminata.*

G. *stylosa.*

G. *mucronulata.*

G. *buxifolia,* A. R. 218, B. R. 443.

G. *collina.*

G. *asperifolia.*

HAKEA *pugioniformis,* B. C. 353.

H. *parilis.*

H. *obliqua.*

H. *gibbosa.*

HAKEA *acicularis.*

H. *suaveolens.*

H. *florida.*

H. *ilicifolia.*

H. *nitida,* B. M. 2246.

H. *amplexicaulis.*

H. *prostrata.*

H. *ceratophylla.*

H. *undulata.*

H. *oleifolia.*

H. *saligna,* A. R. 215.

H. *cinerea.*

H. *dactyloides.*

H. *elliptica.*

LAMBERTIA *formosa,* A. R. 69, B. R. 528, B. C. 80.

XYLOMELUM *pyriforme.*

XYLOPEA *speciosissima,* B. M. 1128.

LOMATIA *silaifolia,* B. M. 1272.

L. *longifolia.*

RHOPALA *dentata* and *sessilifolia.*

BANKSIA *pulchella.*

B. *sphærocarpa.*

B. *nutans.*

B. *ericifolia,* B. M. 738.

B. *spinulosa,* A. R. 457.

B. *collina.*

B. *occidentalis.*

B. *littoralis.*

B. *marginata,* B. C. 61, B. M. 1947.

B. *integrifolia.*

B. *verticillata.*

B. *coccinea.*

B. *paludosa,* B. R. 697, B. C. 392.

B. *oblongifolia,* B. C. 241.

BANKSIA *latifolia.*

B. *marcescens,* A. R. 258.

B. *attenuata.*

B. *serrata,* A. R. 82.

B. *æmula,* B. R. 688.

B. *quercifolia.*

B. *speciosa.*

B. *grandis.*

B. *repens.*

DRYANDRA *floribunda.*

D. *tenuifolia.*

D. *cuneata.*

D. *armata.*

D. *formosa.*

D. *plumosa.*

D. *obtusa.*

D. *nivea.*

D. *longifolia,* B. M. 1582.

AULAX *pinifolia,* A.R. 76, and *umbellata,* A. R. 248.

LEUCADENDRON *argenteum.*

L. *plumosum.*

L. *imbricatum.*

L. *buxifolium.*

L. *Levisanus.*

L. *linifolium.*

L. *stellatum,* B. M. 881.

L. *tortum.*

L. *cinereum.*

L. *corymbosum,* A. R. 495, B. R. 402.

L. *decorum.*

L. *concolor,* A. R. 207.

L. *grandiflorum.*

L. *decurrens.*

L. *strictum.*

LEUCADENDRON *virgatum.*

L. *adscendens.*

L. *concinnum.*

L. *salignum.*

L. *uliginosum.*

L. *floridum,* A. R. 572.

L. *æmulum,* A. R. 429.

L. *abietinum,* A. R. 461.

L. *scabrum.*

BRABEJUM *stellulifolium.*

THYMELEÆ.

PIMELEA *linifolia,* B. M. 891, *rosea,* B. C. 88, B.M. 1458, and *pauciflora,* B. C. 179, heath-looking little shrubs from Australasia, grown in loam and peat, and increased by young cuttings in sand.

STRUTHIOLA *juniperina,* B. C. 75.

S. *erecta,* B. C. 74, B. M. 222.

S. *ovata,* A. R. 119, B. C. 141.

S. *imbricata,* A. R. 113.

S. *tomentosa,* A. R. 334.

S. *virgata,* A. R. 139.

S. *ciliata,* A. R. 149.

S. *pubescens,* B. M. 1212.

Handsome shrubby plants, which grow in sandy loam and peat, and are increased by young cuttings planted in sand and covered with a bell-glass.

DAPHNE *odora,* B. M. 1587, a handsome evergreen shrub from China, which grows in loam and peat, and is best propagated by grafting on any hardy sort.

GNIDIA *pinifolia,* B. R. 19, B. C. 7.

G. *imberbis,* B. M. 1463.

G. *simplex,* B. M. 812.

G. *capitata.*

GNIDIA *oppositifolia*, B. R. 2, B. C. 16.

G. *sericea*, A. R. 225.

Handsome Cape shrubs, which require the same treatment as *Struthiola*.

PASSERINA *filiformis*.

P. *hirsuta*, B. M. 1949.

P. *tenuiflora*.

P. *capitata*.

P. *uniflora*.

P. *grandiflora*, B. M. 292.

P. *spicata*, B. C. 311.

P. *laxa*.

Elegant evergreens, of the same culture as *Struthiola*.

LACHNÆA *conglomerata*.

L. *eriocephala*, B. M. 1295.

L. *purpurea*, B. M. 1594, B. C. 273.

L. *glauca*, B. M. 1658.

L. *buxifolia*, B. M. 1657. Culture as in *Struthiola*.

DAIS *cotinifolia*, B. M. 147, a handsome Cape shrub, which like most others from that country thrives well in loam and peat. It may be increased by cuttings of the roots placed in a warm situation.

SANTALACEÆ.

THESIUM *amplexicaule*, a Cape shrub which grows well in sandy loam and peat, and cuttings root freely under a hand-glass.

FUCHSIA *coccinea*, B. M. 97, *Ilycioides*, B. M. 1024, the first one of the handsomest-flowering shrubs of the green-house, and the second also beautiful. Culture already given, Part I. p. 92.

OSYRIS *alba*, a shrub from the South of Europe, of easy culture and increase in sandy loam.

FUSANUS *compressus,* a Cape shrub of common culture.

ELÆAGNEÆ.

ELÆAGNUS *orientalis,* a Levant shrub of common culture in sandy loam.

ARISTOLOCHEÆ.

ARISTOLOCHIA *glauca,* B. M. 1115, *sempervirens,* B. M. 1116, B. C. 231, *rotunda, hirta,* and *arborescens,* free-growing climbers in loam and peat, which are increased by cuttings in the same soil under a hand-glass.

EUPHORBIACEÆ.

EUPHORBIA *Caput-Medusæ.*
E. *tessellata.*
E. *fructuspina.*
E. *procumbens.*
E. *patula.*
E. *anacantha,* B. C. 220.
E. *clava.*
E. *mauritanica.*
E. *Lamarckii.*
E. *hamata.*
E. *Ornithopus.*
E. *aphylla.*
E. *balsamifera.*
E. *atropurpurea.*
E. *piscatoria.*
E. *mellifera,* B. M. 1305.
E. *prunifolia.*
E. *tuberosa.*

EUPHORBIA *læta.*

E. *genistoides.*

E. *spinosa.*

E. *nummulariæfolia.*

E. *Pithyusa.*

E. *Paratias.*

E. *juncea.*

E. *aleppica.*

E. *serrata.*

E. *spathulæfolia.*

E. *sylvatica.*

E. *imbricata.*

E. *nicæensis.*

E. *Characias.*

These are singular plants, shrubby and herbaceous, and some very unlike others, being low massy succulents with few leaves. These require very little water and a poor sandy soil : the others, and especially the South of Europe species, grow in light rich soil kept moist. Cuttings of most of the species root freely in sand : those of the succulents should be dried before planting, and some of them strike the better for a little bottom heat.

CROTON *maritimum,* a shrubby plant from Carolina, which grows in sandy soil, and is increased by cuttings in sand.

RICINUS *africanus* and *lividus,* Cape shrubs which grow in loam and plenty of leaf-mould, and may be increased either by seeds or cuttings.

STILLINGIA *sylvatica,* a Carolina perennial of easy culture in peat soil.

MERCURIALIS *elliptica,* a Portugal shrub of easy culture in sandy soil.

KIGGELARIA *africana,* a shrub which grows in loam

and peat, and ripened cuttings root in sand under a hand-glass.

Excœcaria *serrata*, a Chili shrub which may be treated as *Kiggelaria*.

Cluytia *alternoides*, B. M. 1321, *polygonoides, daphnoides, ericoides, polifolia, tomentosa, pulchella*, B. M. 1945, *collina*, shrubs from the Cape which grow in loam and peat, and are increased by young cuttings in sand under a bell-glass.

URTICEÆ.

Forskohlea *candida*, a Cape perennial of common culture in loam and peat.

Gunnera *perpensa*, a Cape perennial which grows in sandy loam, and is increased by dividing at the root or by cuttings.

Urtica *nivea*, a Chinese nettle, which grows in light soil and is increased by dividing at the root.

Bœhmeria *rubescens*, a Canary shrub of the easiest culture.

Ficus *cordata, macrophylla, australis, elastica, stipulata, pumila, aspera*, and *capensis*. Trees, shrubs, and creepers or climbers, of the easiest culture and propagation in any light rich soil.

AMENTACEÆ.

Ulmus *chinensis*, a Chinese tree, of easy culture in light rich soil. It may be increased by grafting on any of the hardy species.

Myrica *æthiopica, serrata, laciniata, quercifolia*, and *cordifolia*, evergreen shrubs from the Cape, which thrive in loam and peat, and cuttings root under a hand-glass.

CONIFEREÆ.

Casuarina *equisetifolia*, B. C. 607, *stricta*, A. R. 346, *distyla*, *torulosa*, and *quadrivalvis*.

Curious Australasian trees of little beauty, but valuable where there is room, as they flower from November to February. They grow in loam and peat, and are increased by cuttings in sand under a hand-glass.

Pinus *canariensis*, *longifolia*, and *lanceolata*, evergreen trees from the East, which grow in loam and peat, and cuttings will root in sand under a hand-glass, but not readily. P. *lanceolata* is most easily struck.

Thuja *cupressoides*, an African evergreen shrub, which may be treated like *Pinus*.

T. *juniperoides* and *australis* may be treated in the same manner, but they are oftenest raised by seeds procured from abroad.

Podocarpus *macrophyllus*, *verticillatus*, and *elongatus*, trees or shrubs from the East, which grow in loam and peat, and cuttings root readily in sand under a hand-glass.

Araucaria *imbricata* and *excelsa*, trees from Brazil of the nature of the pine or fir kind. They grow in loam and peat, and may be struck like *Pinus*, but with difficulty.

Juniperus *bermudiana*, *chinensis*, and *barbadensis*, trees from the West Indies which may be treated like *Pinus*.

Taxus *nucifera*, an evergreen low tree from China, of the same culture as *Pinus*.

HYDROCHARIDEÆ.

Trapa *bicornis*, a Chinese aquatic which grows in

loam and peat in shallow water, and is increased by
dividing at the root.

ALISMACEÆ.

TRIGLOCHIN *bulbosum*, B. M. 1445, a marsh plant, of
easy culture in peat soil, set in a saucer of water.

ACTINOCARPUS *minor*, a perennial from New South
Wales of the easiest culture.

SAGITTARIA *sinensis*, B. M. 1631, *lancifolia*, B. M.
1792, and *graminea*, aquatics to be grown in deep water
in any sort of soil, and increased by dividing at the root.

ORCHIDEÆ.

ORCHIS *longicornis*, B. R. 202, and *acuminata*, B. M.
1932, perennials from Barbary, which thrive best in
chalky soil, and when they are in a dormant state they
must have very little water. They are only to be increased by seeds.

BARTHOLINA *pectinata* grows in loam and peat, requires little water when not in a growing state, and is
only to be increased by seeds. As the other *Orchideæ*
require exactly the same treatment, we shall merely enumerate them.

SERAPIAS *Lingua* and *cordigera*, A. R. 475.

OPHRYS *tenthredinifera*, B. R. 205.

SATYRIUM *cucullatum*, A. R. 315, B. C. 104.

S. *carneum*, B. M. 1512.

DISA *cornuta, spathulata*, and *prasinata*, B. R. 210.

PTERYGODIUM *volucris*.

DISPERIS *secunda*.

DIURUS *aurea*.

THELYMITRA *ixioides*.

POGONIA *ophioglossoides.*

CALADENIA *alba.*

GLOSSODIA *major.*

PTEROSTYLIS *obtusa.*

CALEYA *major.*

CALOPOGON *pulchellus,* B. R. 116, B. C. 340.

ARETHUSA *bulbosa.*

DENDROBIUM *speciosum* and *linguiforme.*

These plants, unlike most of the others, may be increased by dividing at the root.

IRIDEÆ.

The culture of this family is the same for every genus, and as it has been already given (Part I. p. 107) need not be here repeated.

TRICHONEMA *Bulbocodium,* B. M. 265.

T. *cruciatum,* B. M. 575.

T. *caulescens,* B. M. 1392.

T. *pudicum,* B. M. 1244.

T. *speciosum,* B. M. 1476.

T. *roseum,* B. M. 1225.

GEISSORHIZA *Rochensis,* B. M. 598.

G. *setacea,* B. M. 1255.

G. *obtusata,* B. M. 672.

G. *secunda,* B. M. 597.

G. *excisa,* B. M. 584.

G. *ciliaris.*

HESPERANTHA *radiata,* B. M. 753.

H. *pilosa,* B. M. 1475.

H. *graminifolia,* B. M. 1254.

H. *falcata,* B. M. 566.

H. *cinnamomea,* B. M. 1054.

SPARAXIS *tricolor* and varieties, B. M. 381.

Sparaxis *bicolor*, B. M. 548.

S. *grandiflora*, B. M. 541, B. R. 258.

S. *bulbifera*, B. M. 545.

Ixia *linearis*, B. M. 570.

I. *capillaris*, B. M. 617.

I. *aulica*, B. M. 1013.

I. *fucata*, B. M. 1379.

I. *patens*, B. M. 522.

I. *leucantha*.

I. *flexuosa*, B. M. 624.

I. *hybrida*, B. M. 127.

I. *conica*, B. M. 539.

I. *monadelpha*, B. M. 607.

I. *curta*, B. M. 1378.

I. *columellaris*, B. M. 630.

I. *maculata*, A. R. 196, B. R. 530.

I. *ovata*, A. R. 23.

I. *viridiflora*, A. R. 29.

I. *ochroleuca*, B. M. 1285.

I. *erecta*, B. M. 623 and 1173.

I. *crateroides*, B. M. 594.

I. *retusa*, B. M. 629.

I. *scillaris*, B. M. 542.

I. *crispa*, B. M. 599.

I. *capitata*, A. R. 159.

Anomatheca *juncea*, B. M. 606.

Tritonia *crispa*, B. M. 678.

T. *viridis*, B. M. 1275.

T. *rosea*, B. M. 1531.

T. *capensis*, B. M. 618.

T. *longiflora*, B. M. 256.

T. *tenuiflora*, B. M. 1502.

T. *concolor*, B. M. 1502.

T. *Rochensis*, 1503.

TRITONIA *pallida.*

T. *lineata,* B. M. 487.

T. *securigera,* B. M. 383.

T. *flava,* B. C. 747.

T. *squalida,* B. M. 581.

T. *fenestrata,* B. M.704.

T. *crocata,* B. M. 184.

T. *deusta,* B. M. 622.

T. *miniata,* B. M. 609.

T. *refracta,* B. R. 135.

WATSONIA *spicata,* B. M. 523.

W. *plantaginea,* B. M. 553.

W. *punctata,* A. R. 177.

W. *roseo alba,* B. M. 537 and 1193.

W. *marginata,* B. M. 608 and 1530.

W. *strictiflora,* B. M. 1406.

W. *rosea,* B. M. 1072.

W. *brevifolia,* B. M. 601.

W. *iridifolia,* B. M. 600.

W. *fulgida,* B. M. 600.

W. *Meriana,* B.M. 418 and 1194.

W. *humilis,* B. M. 631 and 1195.

W. *aletroides,* 441 and 533.

GLADIOLUS *Cunonia,* B. M. 343.

G. *Watsonius,* B. M. 450 and 569.

G. *quadrangularis,* B. M. 567.

G. *viperatus,* B. M. 688.

G. *alatus,* B. M. 586.

G. *namaquensis,* B. M. 592.

G. *brevifolius,* B. M. 727 and 992.

G. *hirsutus,* B. M. 574.

G. *versicolor,* B. M. a. 1042.

G. *binervis,* B. M. e. 1042.

G. *edulis,* B. R. 169.

GLADIOLUS *hastatus,* B. M. 1564.

G. *tristis,* B. M. 272.

G. *concolor,* B. M. 1098.

G. *trichonemifolius,* B. M. 1483.

G. *gracilis,* B. M. 562.

G. *recurvis,* B. M. 578.

G. *carneus,* B. M. 591.

G. *cuspidatus,* B. M. 582.

G. *blandus,* B. M. 625.

G. *campanulatus,* A. R. 188.

G. *augustus,* B. M. 602.

G. *involutus.*

G. *undulatus,* B. M. 538.

G. *floribundus,* B. M. 610.

G. *Milleri,* B. M. 632.

G. *cardinalis,* B. M. 135.

G. *Byzantinus,* B· M. 874.

G. *communis,* B. M. 86.

MELASPHÆRULA *graminea,* B. M. 615.

ANTHOLYZA *vittigera,* B. M. 1172 and *æthiopica,* B. M. 561.

BABIANA *Thunbergii.*

B. *ringens.*

B. *tubiflora,* B. M. 847.

B. *tubata,* B. M. 680.

B. *spathacea,* B. M. 638.

B. *sambucina,* B. M. 1019.

B. *disticha,* B. M. 626.

B. *plicata,* B. M. 576.

B. *stricta,* B. M. 621 and 637.

B. *purpurea,* B. M. 1052.

B. *villosa,* B. M. 583.

B. *rubro cyanca,* B. M. 410.

ARISTEA *cyanea,* B. M. 458.

Aristea *capitata*, B. M. 605.

A. *spiralis*, B. M. 520.

A. *melaleuca*, B. M. 1277.

A. *pusilla*, B. M. 1231.

Witsenia *maura*, B. R. 5, and *corymbosa*, B. M. 895, B. C. 254.

Lapeyrousia *corymbosa*, B. M. 595, and *fissifolia*, B. M. 1246.

Moræa *flexuosa*, B. M. 695.

M. *collina*, B. M. 1033.

M. *miniata*, A. R. 404.

M. *pavonia*, B. M. 1247.

M. *tripetala*, B. M. 702.

M. *angusta*, B. M. 1276.

M. *tricuspis*, B. M. 696 and 772.

M. *tenuis*, B. M. 1047.

M. *unguiculata*, B. M. 593.

M. *edulis*, B. M. 613 and 1238.

M. *longiflora*, B. M. 712.

M. *spicata*, B. M. 1283.

M. *tristis*, B. M. 577.

M. *crispa*, B. M. 1284.

M. *bituminosa*, B. M. 1045.

M. *viscaria*, B. M. 587.

M. *ramosa*, B. M. 771.

M. *villosa*, B. M. 571.

M. *barbigera*, B. M. 1012.

M. *ciliata*, B. M. 1061.

M. *Sisyrinchium*, B. M. 1407.

M. *papilionacea*, B. M. 750.

M. *spathacea*.

Iris *moræoides*, B. M. 693.

Marica *gladiata*, B. R. 229, and M. *californica*, B. M. 983.

Patersonia *sericea* and *glabrata*, B. M. 1041 and 50.

FERRARIA *undulata*, B. M. 144, and *antherosa*, B. M. 751.

GALAXIA *ovata*, A. R. 94, B. M. 1208, *grandiflora*, A. R. 164, *mucronularis*, *versicolor*, and *graminea*, B. M. 1292.

HÆMODORACEÆ.

WACHENDORFIA *thyrsiflora*, B. M. 1060.

W. *paniculata*, B. M. 616.

W. *graminea*.

W. *hirsuta*, B. M. 614.

W. *brevifolia*, B. M. 1166.

Bulbs which may be treated like the *Irideæ*.

DILATRIS *corymbosa* and *viscosa*. Bulbs of the same nature as those of *Irideæ*.

HÆMODORUM *planifolium*, B. M. 1610, a New Holland perennial, which grows chiefly in loam and peat, and is increased by dividing at the roots.

LANARIA *plumosa* may be treated like *Hæmodorum*.

ANIGOZANTHUS *flavida*, B. M. 1151, may be treated like *Hæmodorum*, with rather more water when in a growing state.

AMARYLLIDEÆ.

HÆMANTHUS *coccineus*, B. M. 1075, B. C. 240.

H. *coarctatus*, B. R. 181.

H. *rotundifolius*, B. M. 1618.

H. *puniceus*, B. M. 1315.

H. *albiflos*, B. M. 1239, B. C. 602.

H. *tigrinus*, B. M. 1705.

H. *quadrivalvis*, B. M. 1523.

H. *pubescens*, B. R. 382.

HÆMANTHUS *maculatus*.

H. *lanceæfolius*.

H. *carinatus*.

H. *Pumilio*.

Cape bulbs which grow in sandy loam with a little peat; they require no water when not in a growing state, and are increased by offsets.

STRUMARIA *truncata*.

S. *rubella*.

S. *angustifolia*.

S. *linguæfolia*.

S. *filifolia*.

S. *spiralis*, B. M. 1383.

S. *crispa*, B. M. 1363.

S. *stellaris*.

S. *gemmata*, B. M. 1620.

CRINUM *pedunculatum*, B. R. 52.

CYRTANTHUS *angustifolius*, B. M. 271.

C. *collinus*, B. R. 162.

C. *spiralis*, B. R. 167.

C. *obliquus*, B. M. 1133.

C. *uniflorus*, B. R. 168.

BRUNSVIGIA *Josephinæ*, and J. *minor*, B. R. 192 and 193.

B. *multiflora*, B. M. 1619.

B. *marginata*.

B. *Radula*.

B. *striata*.

B. *falcata*, B. M. 1443, B. C. 745.

B. *toxicaria*, B. M. 1217, B. R. 567.

B. *ciliaris*.

A beautiful family of bulbs, some of which grow to a great size, and require large pots to have them flower in perfection. They grow in sandy loam with a little peat;

and require ample supplies of water when in a growing state, but very little when dormant. They are increased by offsets or seeds.

AMARYLLIS *chloroleuca.*

A. *Pumilio.*

A. *perdica.*

A. *formosissima,* B. M. 47.

A. *blanda,* B. M. 1450.

A. *Belladonna pallida,* B. M. 733. B. R. 714.

A. *vittata,* B. M. 129.

A. *purpurea,* B. M. 1430, B. R. 552.

A. *coronica,* B. R. 139.

A. *aurea,* B. M. 409, B. R. 611.

A. *curvifolia,* B. M. 725.

A. *corusca,* B. M. 1089.

A. *sarniensis,* B. M. 294.

A. *venusta,* B. M. 1090.

A. *radiata,* A. R. 95, B. R. 596.

A. *undulata,* B. M. 369.

A. *humilis,* B. M. 726.

A. *flexuosa,* B. R. 172.

A. *longifolia,* B. M. 661, B. R. 303.

A. *revoluta,* B. M. 915 and 1178, B. R. 615.

This genus thrives best in a rich loamy soil, and with spare waterings after having done flowering. The species are increased by offsets; or " a shell or coat taken from the bulb with a leaf on it, and planted in a pot of sandy soil, will produce a bulb, as will almost any bulbous-rooted plant." (*Bot. Cult.* 131.)

PANCRATIUM *carolinianum* and *canariense,* B. R. 174, bulbs which grow in light loam, and are increased by offsets and seeds.

DORYANTHES *excelsa,* B. M. 1685, an elegant plant which grows in sandy loam and peat, but does not flower till it gets large, and should therefore be planted in a conservatory. It is increased by suckers.

GETHYLLIS *spiralis*, B. M. 1088, *ciliaris*, *villosa*, and *lanceolata*, Cape bulbs which grow in sandy loam and peat, and are otherwise of common bulb culture.

BLANDFORDIA *nobilis* and *grandiflora*, beautiful plants which grow in sandy loam and peat, and are increased by suckers or seeds.

AGAPANTHUS *umbellatus*, B. M. 500, B. R. 699, and *præcox*, handsome plants which thrive in loam with a little rotten dung, and are increased by dividing at the root.

TRITOMA *uvaria*, B. M. 758.

T. *media*, B. M. 744.

T. *pumila*, B. M. 764, B. C. 444.

Frame plants which will grow in any light soil, and are increased by dividing at the root.

VELTHEIMIA *viridifolia*, B. M. 501, and *glauca*, B. M. 1091, Cape bulbs which grow in light loamy soil, and are increased by offsets or leaves.

POLYANTHES *tuberosa*, B. R. 63 (see P. I. page 112).

TULBAGIA *alliacea*, B. M. 806, and *cepacea*, Cape bulbs which grow in sandy loam and peat, and are increased by offsets or seeds.

BRODIÆA *grandiflora* and *congesta*, American herbaceous plants of the usual culture in loam and peat.

ALOE *purpurascens*, B. M. 1474.

A. *soccotrina*, B. M. 472.

A. *arborescens*, B. M. 1306.

A. *ferox*, B. M. 1975.

A. *supralævis*.

A. *flavispina*.

A. *picta*, B. M. 1323.

A. *latifolia*, B. M. 1346.

A. *saponaria*, B. M. 1460.

ALOE *serrulata*, B. M. 1415.

A. *mitræformis*, B. M. 1270.

A. *nobilis*.

A. *distans*.

A. *brevifolia*, B. M. 1362.

A. *depressa*, B. M. 1332.

A. *suberecta*.

A. *acuminata*, B. M. 1322.

A. *tuberculata*.

A. *humilis*, B. M. 757.

A. *incurva*, B. M. 828.

A. *striata*.

A. *glauca*, B. M. 1278.

A. *africana*.

A. *virens*, B. M. 1355.

A. *plicatilis*, B. M. 457.

A. *variegata*, B. M. 513.

A. *nigricans*, B. M. 838.

A. *glabra*.

A. *carinata*, B. M. 1331.

A. *Lingua*, B. M. 979.

A. *angulata*.

A. *angustifolia*.

A. *longifola*.

A. *obtusifolia*.

A. *intermedia*.

A. *verrucosa*, B. M. 837.

A. *spiralis*, B. M. 1455.

A. *spirella*.

A. *pentagona*, B. M. 1338.

A. *imbricata*, B. M. 1455.

A. *foliolosa*, B. M. 1352.

A. *tortuosa*.

A. *rigida*, B. M. 1337.

A. *viscosa*, B. M. 814.

Aloe *cymbiformis*, B. M. 802.

A. *reticulata*, B. M. 1314.

A. *recurva*, B. M. 1353.

A. *recusa*, B. M. 455.

A. *mirabilis*, B. M. 1354.

A. *pumila*, B. M. 1361.

A. *Radula*, B. M. 1345.

A. *attenuata*, B. M. 1345.

A. *minima*, B. M. 1360.

A. *minor*, B. M. 815.

A. *margaritifera*.

A. *Haworthii*.

These are succulents from the Cape, most of them of very humble growth, and all of them singular in appearance. They grow best in sandy loam mixed with a little lime rubbish or gravel, and flower the better for being exposed to the open air in summer. They are increased by suckers, or cuttings of the shrubby kinds, or leaves stripped off the plants and laid on a pot of mould, or planted shallow in it, will produce young plants.

SMILACEÆ.

Ophiopogon *japonicus*, B. M. 1063, a frame perennial from Japan, which grows in light rich soil, and is increased by dividing at the root.

Myrsiphyllum *asparagoides* and *angustifolium*, herbaceous climbers which grow in sandy loam and peat, and are increased by dividing at the root.

Trillium *pumilum*, a perennial from Carolina, which grows in loam and peat, and is increased by dividing at the root.

Smilax *excelsa, China, australis, latifolia, glyciphylla* (Botany-bay Tea), and *Pseudo-China*, climbing shrubs which grow freely in loam and peat, and young cuttings root in sand under a hand-glass.

TAMUS *elephantipes*, B. M. 1347, a curious Cape plant, which grows in rich light soil, and must not be over-watered. It is rare, and has not yet been propagated; but Mr. Sweet is of opinion it will increase by cuttings in sand in bottom heat.

RUSCUS *androgynus*, B. M. 1898, a climbing shrub from the Canaries, which grows in rich light soil, and is increased by dividing at the root.

LILIACEÆ.

FRITILLARIA *lutea*, B. M. 1538, a bulb which grows in sandy loam, and is increased in the usual manner.

YUCCA *aloifolia*, B. M. 1700, *serrulata* and *draconis*, evergreen shrubs with large long leaves, which thrive in rich loamy soil, and are increased by suckers.

BROMELIACEÆ.

AGAVE *americana*, A. R. 438, *Milleri*, *angustifolia*, *virginica*, B. M. 1157, and *gemmiflora*, succulent evergreens, which thrive well in rich loamy soil, and are increased by suckers.

ASPHODELEÆ.

EUCOMIS *nana*, B. M. 1495, *purpureocaulis*, A.R. 369, *bifolia*, B. M. 840, *regia*, *undulata*, B. M. 1083, *punctata*, B. M. 913, and *striata*, B. M. 1539, Cape bulbs, which grow in any rich light soil, and are increased by offsets or seeds, or by leaves.

SOWERBÆA *juncea*, B. M. 1104, a New Holland perennial, which grows in peat soil kept moist, and is increased by dividing at the root.

ALLIUM *Chamæ-Moly*, B.M. 1203, a bulb of common culture and propagation in any light rich soil.

ALBUCA *altissima.*

A. *major*, B. M. 804.

A. *minor*, B. M. 720.

A. *flaccida.*

A. *viridiflora*, B. M. 1656.

A. *coarctata.*

A. *caudata.*

A. *setosa*, B. M. 1481.

A. *vittata*, B. M. 1329.

A. *physodcs*, B. M. 1046.

A. *exuviata*, B. M. 871.

A. *anthericoides.*

A. *fragrans.*

A. *viscosa.*

A. *spiralis.*

A. *fastigiata*, A. R. 480, B. R. 277.

Bulbs which thrive in sandy loam and leaf-mould, and are increased by offsets or leaves.

XANTHORRŒA *Hastile, minor* and *bracteata*, perennials which grow in loam and peat, and are increased by offsets from the roots.

THYSANOTUS *junceus* grows freely in loam and peat, and is increased by dividing at the roots.

ERIOSPERMUM *latifolium*, B. M. 1382, *lanceæfolium*, *parvifolium* and *folioliferum*, A. R. 521, tuberous-rooted plants which grow in loam and peat, and are increased by offsets or seeds.

ORNITHOGALUM *niveum*, B. R. 235.

O. *lacteum*, B. M. 1134.

O. *revolutum*, B. M. 653, B. R. 315.

ORNITHOGALUM *elatum,* A. R. 528.

O. *latifolium,* B. M. 876.

O. *scilloides.*

O. *prasinum,* B. R. 158.

O. *odoratum,* A. R. 260.

O. *barbatum.*

O. *juncifolium,* B. M. 972.

O. *rupestre.*

O. *arabicum,* B. M. 728.

O. *thyrsoides,* B. M. 1164.

O. *aureum,* B. M. 190.

O. *flavissimum.*

O. *coarctatum.*

O. *caudatum,* B. M. 805.

O. *unifolium,* B. M. 953 and 935.

O. *Squilla,* B. M. 918.

Bulbs which grow in sandy loam and peat, and are increased by offsets or seeds.

SCILLA *brevifolia,* B. M. 1468, requires the same treatment as *Ornithogalum.*

MASSONIA *latifolia,* B. M. 848.

M. *muricata,* B. M. 559.

M. *pustulata,* B. M. 642.

M. *echinata.*

M. *pauciflora.*

M. *angustifolia,* B. M. 736.

M. *undulata.*

M. *ensifolia,* B. M. 554.

Cape bulbs which grow in sandy loam and peat, and require to be shifted annually when just beginning to grow; water must be sparingly supplied till they have made abundance of roots, and must be totally withheld when they are in a dormant state. They are increased by offsets and seeds.

ANTHERICUM *frutescens,* B. M. 816.
A. *rostratum.*
A. *alooides.*
A. *nutans.*
A. *latifolium.*
A. *pugioniforme,* B.M. 1454.
A. *asphodeloides.*
A. *longiscapum,* B. M. 1339.
A. *annuum,* B.M. 1451.
A. *hispidum.*
A. *fragrans.*
A. *flexifolium.*
A. *filiforme.*
A. *floribundum.*
A. *revolutum,* B. M. 1044.
A. *vespertinum,* B. M. 1040.
A. *graminifolium.*
A. *triflorum.*
A. *canaliculatum,* B. M. 1124.

Shrubby, succulent, bulbous and annual plants, which grow in light sandy soil with the pots well drained, and are increased by cuttings, offsets, dividing at the root, or by seeds.

ARTHROPODIUM *paniculatum,* B. M. 1421, may be treated like *Eriospermum.*

CHLOROPHYTUM *elatum* grows in loam and leaf-mould, and is increased by offsets or seeds.

CÆSIA *vittata* may be treated like *Chlorophytum.*

DIANELLA *cærulea,* B. M. 505, and *divaricata,* plants from New South Wales, which grow in loam and peat and are increased by dividing at the root.

EUSTREPHUS *latifolius,* B. M. 1245, a New Holland climber, which is increased by dividing at the root or by cuttings. It grows in loam and peat with sand, and flowers in June and July.

Asparagus *declinatus.*

A. *decumbens.*

A. *scandens.*

A. *retrofractus.*

A. *asiaticus.*

A. *æthiopicus.*

A. *albus.*

A. *acutifolius.*

A. *flexuosus.*

A. *aphyllus.*

A. *subulatus.*

A. *capensis.*

Shrubs, chiefly evergreens and climbers, which grow in sandy loam, and are increased by dividing at the root, or cuttings under a hand-glass without bottom heat.

Drimia *altissima,* B. M. 1074.

D. *elata,* B. M. 822.

D. *ciliaris,* B. M. 1444.

D. *pusilla.*

D. *lanceæfolia,* B. M. 643, B. C. 278.

D. *revoluta,* B. M. 1380.

Cape bulbs which grow in sandy loam and leaf-mould, require to be fresh potted just when they are beginning to grow, and are increased by offsets or seeds.

Uropetalon *viride.*

U. *glaucum,* B. R. 156.

U. *crispum.*

U. *serotinum,* B. M. 859.

U. *fulvum,* B. M. 1185.

Lachenalia *glaucina.*

L. *orchioides.* B. M. 1269.

L. *pallida,* B. R. 287, 314.

LACHENALIA *hyacinthoides.*

L. *angustifolia*, B. M. 735.

L. *contaminata*, B. M. 1401.

L. *patula.*

L. *fragrans.*

L. *unicolor*, B. M. 1373.

L. *lucida*, B. M. 1372.

L. *racemosa*, B. M. 1517.

L. *pustulata*, B. M. 817.

L. *purpureo-cærulea*, B. M. 745.

L. *nervosa*, B. M. 1497.

L. *violacea.*

L. *bifolia*, B. M. 1611.

L. *rosea*, A. R. 296.

L. *unifolia*, B. M. 766.

L. *sessiliflora*, A. R. 460.

L. *isopetala.*

L. *tricolor*, B. M. 89.

L. *luteola.*

L. *quadricolor*, B. M. 588 and 1097.

L. *rubida*, B. M. 993.

L. *pendula*, B. M. 590, B. C. 267.

Cape bulbs of low growth, which grow in loam and leaf-mould, require little water when not in a growing state, and are increased by offsets or seeds.

ALSTRŒMERIA *Pelegrina*, B. M. 139; a fine plant which grows in loam and leaf-mould, and is increased by dividing at the root or by seeds, which as they soon lose their vegetative power should be sown when ripe.

PHORMIUM *tenax* thrives in any rich light soil, and is increased by offsets from the root.

HYPOXIS *sobolifera*, B. M. 711.

H. *villosa.*

H. *decumbens.*

H. *obliqua*, A. R. 195.

HYPOXIS *aquatica.*

H. *alba.*

H. *obtusa*, B. R. 159.

H. *ovata*, B. M. 1010.

H. *stellata*, B. M. 662.

H. ———— *elegans*, B. M. 1223.

H. *veratrifolia.*

H. *linearis*, A. R. 171.

H. *serrata*, B. M. 709.

H. *juncea.*

Cape bulbs which may be treated like *Lachenalia.*

CURCULIGO *plicata*, B. R. 345, may also be treated like *Lachenalia.*

CYANELLA *capensis*, B. M. 568, B. C. 732, and *lutea*, B. M. 1252. See *Lachenalia.*

MELANTHACEÆ.

MELANTHIUM *monopetalum*, B. M. 1291.

M. *spicatum*, B. M. 694.

M. *pumilum.*

M. *longiflorum.*

M. *junceum*, B. M. 558.

M. *secundum.*

M. *capense.*

M. *uniflorum*, B. M. 767.

M. *eucomoides*, B. M. 641.

M. *viride*, B. M. 994.

Cape bulbs which thrive in loam and peat, and are increased by offsets or seeds.

RESTIACEÆ.

XYRIS *operculata*, B. M. 1158, B. C. 205, thrives in peat soil, and is increased by dividing at the root.

Willdenovia *teres* may be treated as *Xyris.*

Restio *tectorum* is of the same culture as *Xyris.*

Elegia *juncea,* and *racemosa,* may be treated like *Xyris.*

COMMELINEÆ.

Commelina *africana,* B. M. 1431, grows in sandy loam, and is increased by dividing at the root or by seeds.

CANNEÆ.

Canna *patens,* B. R. 576, grows in loam and peat, and is increased by dividing at the root or by seeds.

Thalia *dealbata,* B. M. 1690, grows in loam and peat, either in or out of a cistern of water, and is increased like *Canna.*

SCITAMINEÆ.

Zingiber *Mioga,* Japanese Ginger, may be treated like *Canna.*

CYPERACEÆ.

All the plants of this Genus are of the grassy or sedgy kind, and grow in loamy soil or loam and peat, and are increased by dividing at the root or by seeds. They are plants of no beauty whatever.

Cyperus *elegans, alopecuroides,* and *badius.*

Carex *appressa.*

AROIDEÆ.

Calla *æthiopica* may be treated as an aquatic, and

grown in deep water in pots of rich loam, or it will grow in loam and peat on the common stage of the green-house. It is increased by suckers.

ARUM *crinitum*, *ternatum*, and *Arisarum*, frame herbaceous plants, which grow in sandy loam, and are increased by suckers or dividing at the root.

GRAMINEÆ.

All the Grasses grow in any common soil, and are increased by dividing at the root or by seeds.

PASPALUM *stoloniferum*.

EHRHARTIA *panicca*.

ANDROPOGON *muticus*.

NAIADEÆ.

APONOGETON *distachyon*, B. M. 1293, and *angustifolium*, B. M. 1268, aquatic bulbs which grow in loam and peat plunged in a cistern, and are increased by offsets or seeds.

FELICEÆ.

All the Ferns grow in loam and peat, and are increased by dividing at the root or by seeds.

ACROSTICHUM *Lingua*, *alcicorne*, B.R. 262-3, *velleum*.

NEPHRODIUM *elongatum*.

ALLANTODIA *umbrosa*.

ASPIDIUM *axillare* and *æmulum*.

ASPLENIUM *monanthemum* and *rhizophyllum*.

PTERIS *arguta* and *esculenta*.

BLECHNUM *australe*.

WOODWARDIA *radicans*.

DOODIA *aspera*, B. C. 39.

ADIANTUM *reniforme.*

CHEILANTHES *pteroides* and *fragrans.*

DAVALLIA *pyxidata* and *canariensis,* B. C. 142

DICKSONIA *Culcita.*

GENERA WHOSE NATURAL ORDERS ARE NOT YET GENERALLY KNOWN.

POLLICHIA *campestris,* a biennial which grows in any light rich soil, and is increased by cuttings in the same soil or by seeds.

PENÆA *mucronata* and *squamosa,* B. R. 106, handsome Cape shrubs which grow in sandy loam and peat, and cuttings root in sand under a bell-glass.

CURTISIA *faginea,* a Cape tree with broad leaves : it grows in loam and peat, and ripened cuttings root in sand under a bell glass.

CHLORANTHUS *inconspicuus* and *erectus,* shrubs from China and the Cape, which are of the same culture as *Curtisia.*

LASIOPETALUM *parviflorum,* B. C. 413.

L. *ferrugineum,* B. M. 1766.

L. *purpureum,* B. M. 1755, B. C. 361.

L. *arborescens.*

L. *solanaceum,* B. M. 1486, B. C. 279.

L. *quercifolium,* B. M. 1485, B. C. 619.

New Holland shrubs which grow in sandy loam and a little leaf-mould, and ripened cuttings root readily in sandy loam under a hand-glass.

CALODENDRUM *capense,* one of the finest trees known, with a fruit like a chesnut : it grows and is propagated like *Lasiopetalum.*

COMMERSONIA *dasyphylla,* A. R. 603, a New Holland shrub which may be treated like *Lasiopetalum.*

NANDINA *domestica*, B. M. 1109, grows in loam and peat, and young cuttings root in sand under a bell-glass.

BORONIA *pinnata*, B. C. 473, and *serrulata*, grow in sandy peat, and are increased by layers or by cuttings, care being taken to wipe the glasses to prevent their damping off.

TETRATHECA *juncea* grows in loam and peat, and is increased by young cuttings in sand under a bell-glass.

CROWEA *saligna*, B. M. 989, B. C. 310, grows in sandy peat and loam, and is increased by young cuttings in the same soil.

CODON *Royeni*, A. R. 325, a Cape biennial of common culture.

DIONÆA *Muscipula*, B.M. 785: this plant thrives best when planted in a pot of moss with a little earth at the bottom, and the pot placed in a pan of water: leaves slipped off and planted in moist moss will root and become plants.

SARRACENIA *flava*, B. M. 780, *variolaris*, B. M. 1710, *rubra* and *purpurea*, B. M. 849, frame plants from North America, which may be treated like *Dionæa*.

HYÆNANCHE *globosa*, a Cape shrub which grows in loam and peat, and is increased by young cuttings in sand under a bell-glass.

EUCLEA *racemosa* and *undulata*, shrubs which grow in loam and peat, and ripened cuttings root readily in sandy loam under a bell- or hand-glass.

HAMILTONIA *oleifera*, an American shrub which grows in peat soil, and is increased by young cuttings in sand under a bell-glass.

LAUROPHYLLUS *capensis*, a Cape shrub which may be treated as *Hamiltonia*.

THE END.

INDEX.

The roman numerals in the preceding Index refer to the First or Second Parts of the Work respectively.

Printed by Richard Taylor,
Shoe-lane, London.

Printed in the United States
By Bookmasters